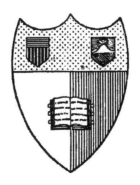

Cornell University
Ithaca, New York

COLLEGE OF ARCHITECTURE
LIBRARY

HOW TO PLAN
THE HOME GROUNDS

How to Plan
the Home Grounds

By

S. PARSONS, Jr.

Ex-Superintendent of Parks, New York City
Fellow of the American Society of Landscape Architects

With Illustrations

drawn by W. E. Spader under the direction of the
author and of G. F. Pentecost, Jr., F.A.S.L.A.

NEW YORK
DOUBLEDAY, PAGE & CO.
1905

To the Memory of

CALVERT VAUX AND WILLIAM A. STILES

WHOSE WORDS AND DEEDS HAVE BEEN THE CHIEF INSPIRATION OF THIS
BOOK, AND WHOSE SINGULARLY FELICITOUS EXPRESSION OF THEIR
OWN PERSONAL FORCE AND CHARM SERVED EVER BUT TO
EMPHASIZE THE FEW, SIMPLE UNDERLYING TRUTHS UPON
WHICH ARE BASED ALL MANIFESTATIONS OF THE
ART OF LANDSCAPE GARDENING, FROM THE
MOST RESTRICTED TO THOSE OFFERING
THE LARGEST POSSIBILITIES.

PREFACE

THE purpose of this book is to set forth briefly some simple basic principles concerning the processes whereby home grounds can be made beautiful. From the beginning it will follow the various stages through which may be gradually and naturally developed the sensible, which is always the pleasing and attractive, dwelling place; for everything which is done according to sound rational principles and common sense is bound to be agreeable and beautiful. In as short a fashion and as clearly as lies in the author's power, it will seek to set down the few points which are to be kept always in mind to properly work out and accomplish the permanently satisfactory result. To make these points tangible, by giving with the reason the example which makes that reason evident, the author invariably reverts to the general principles that should never be lost sight of in the selection and arrangement of the territory intended for occupation. These principles apply invariably to the small as well as the large places. The statement cannot be made too emphatically at the very outset, that it is always just as simple and just as difficult to lay out a small yard 25 x 100 feet as a gentleman's great country place of many acres. There may be more details in the large place, but the principles are the same in both, and in the village lot the dainty finish and the perfect proportion, where all things are so evident, may be more difficult to accomplish than the more massive and less emphasized effects of the regular country place.

It is needless to dwell upon the necessity of having entirely good reasons to control the choice of a place. Should that be left to unrestrained fancy or whim, the result is sure to prove disastrous. Why the house should stand in one place and not in another, is not and cannot be a matter of fancy. Certain reasons govern it. Many points have always to be considered. Everything is interdependent. There should be a general scheme from which everything naturally develops in its relative and just order and place, and the basis of all design and of all arrangements should be the natural conformation and incidents of the ground.

It will be evident and natural that the existing landscape must control the general design, but, of course, never to the point where danger threatens the actual comfort of the householder by shutting out sunlight and air, and otherwise making unhealthy and uncomfortable conditions, such as low, damp ground and bleak exposure, for art and beauty in such cases always go hand in hand with common sense and reasonable comfort. The arrangement and construction of roads and paths, whether on large or small places, will then develop in a natural fashion that will be practical and agreeable. Gardens will be seen in their proper places and performing their true functions under existing conditions. In the same spirit, the location of ponds and streams and their construction will be studied, as well as the health of woodlands and the retention of their most characteristic beauty. The natural functions and adjustment of fences, bridges, and summer-houses will also receive attention.

In no way does the author intend to advocate a special style, whether Italian, Colonial, or what not, but only such an arrangement as naturally grows out of the peculiar conformation of the ground under consideration.

The list of the best trees and shrubs, one of the most important features of the book, will be limited in number of kinds discussed, because it is desirable to give it the most general application possible, and the consideration of rare species will be left out, as their employment would naturally involve fancy expenditure and oftentimes difficult and expensive care.

Finally, in illustration of the proper employment of the principles set down and the trees and shrubs discussed, practical diagrams relating to the arrangement of home grounds will be given.

The latter part of the book will be occupied with a brief consideration of Parks, Cemeteries, and Railroad Stations, as typical examples of the more extended development of the principles on which home grounds are naturally laid out.

TABLE OF CONTENTS

PART I

	PAGE
Preface	vii
The Selection of Home Grounds	1
The Selection of the Site of the House	10
Roads and Paths	18
Lawns	45
Flower Gardens	53
The Terrace	70
Plantations	79
Deciduous Trees	94
Deciduous Shrubs	107
Evergreen Trees	121
Evergreen Shrubs	124
Hardy Herbaceous Plants	129
Aquatic Plants	138
Hardy Vines and Climbers	141
Bedding Plants	145
Pools and Streams	151
Woodlands	159
The Use of Rocks	165
Residential Parks	174
Fences, Bridges, and Summer-houses	184
List of Plants for General Use on Home Grounds	195
Contracts and Specifications	204

PART II

	PAGE
PARKS AND PARKWAYS	219
CHURCH-YARDS AND CEMETERIES	225
SEASIDE LAWNS	229
CITY AND VILLAGE SQUARES	233
RAILROAD STATION GROUNDS	240

LIST OF ILLUSTRATIONS

	PAGE
VILLAGE LOT OF HALF AN ACRE, SHOWING SUITABLE LOCATION FOR SMALL HOUSE	12
LATTICE AND VINE PROTECTION FOR DRYING-GROUND	13
CONTOUR MAP OF COUNTRY PLACE OF EIGHT ACRES, WITH STEEP CONTOURS	20
SAME COUNTRY PLACE, SHOWING ARRANGEMENT OF PLANTATIONS, ROADS, AND PATHS	21
SECTION OF SAME PLAN TAKEN ON LINE ACROSS THE CENTER OF CIRCLE, HOUSE, AND TERRACE	21
DOUBLE ENTRANCE, WITH ANTE-PARK	22
CARRIAGE TURN WITH GRASS PLOT, SHOWING COURSE PARALLEL WITH HOUSE	23
BAD LINE FOR A ROAD; GOOD LINE FOR A ROAD	24
BRANCHING ROADS IN LARGE PLACE, ENTERING AT RIGHT ANGLES TO HIGHWAY	25
OPEN CARRIAGE TURN ON MEDIUM-SIZED PLACE	26
OPEN CARRIAGE TURN FOR SMALL PLACE	27
VILLAGE LOT, EIGHT-TENTHS OF AN ACRE, WITH OUTBUILDINGS AND STRAIGHT WALK SYSTEM	29
CROSS SECTION OF ROAD WITH SOD GUTTERS	34
VILLAGE LOT, ONE-HALF OF AN ACRE, WITH STRAIGHT WALKS	40
FORMAL ENTRANCE OF A LARGE PLACE	42
TREATMENT OF ENTRANCE GATE AND LODGE	43

LIST OF ILLUSTRATIONS

	PAGE
Village Lot, One-third of an Acre, with Open Lawn, Overlooking Fine Southwest View	47
Contour Map of Same Village Lot, One-third of an Acre	48
Example of Formal Garden, with Grass Walks	59
Corner of a Formal Flower Garden	60
Small Flower Garden, near House, with Straight Beds and Grass Walks	63
Small Flower Garden, Rear of House, with Elliptical Beds Arranged with Gravel Main Walks and Subsidiary Grass Walks	65
Flower Garden with Bordering Beds, Interior Grass Plot, and Shade Trees at Intersections of Gravel Walks	68
Bird's-eye View of a Terrace on Crest of Hill, with Background of Woods	71
Ground Plan of Same Terrace	72
Cross Section of Terrace Shown on Page 71	75
Pergola on Highest Point of Same Terrace on Page 71	76
Village Corner Lot, Five-twenty-eighths of an Acre, with Bordering Shrubs and Flower Garden	80
Type of Tree Dimension for Elms, Maples, Lindens, Oriental Planes, and Ashes, Suitable for Quickest Effect and Healthy Growth	85
Village Lot, One-half an Acre, Located on Stream	155
Section of Same	155
Summer-house on Water	157
Treatment of Rocks for Stream and Bridge	167
Bridge of Boulders	169
Bridge of Boulders, with Rock Treatment of Stream, in Central Park, New York	170
Rough Stone Wall and Coping	171
Treatment of Steps with Rocks, Central Park, New York	172

LIST OF ILLUSTRATIONS

	PAGE
ALBEMARLE PARK, ASHEVILLE, N. C., SHOWING ARRANGEMENT OF ROADS AND HOUSE LOTS	180, 181
CROSS SECTION OF ALBEMARLE PARK, ASHEVILLE, N. C.	181
LONGITUDINAL SECTION OF ALBEMARLE PARK, ASHEVILLE, N. C.	182, 183
HA-HA FENCE, FOR SEPARATING PLEASURE GROUNDS FROM FARM LANDS	186
IRON PIPE AND ANCHOR-POST FENCE	187
SUMMER-HOUSE, CENTRAL PARK, NEW YORK	189
SUMMER-HOUSE	190
SUMMER-HOUSE, CENTRAL PARK, NEW YORK	191
PERGOLA, OR OPEN VINE-COVERED ARBOR	192
PLAIN RUSTIC BRIDGE IN GENTLEMAN'S COUNTRY PLACE	193
BOW BRIDGE, CENTRAL PARK, NEW YORK	194
STONE BRIDGE OVER SMALL STREAM	194
RURAL PARK OF MODERATE DIMENSIONS FOR CITY OR TOWN	220
DESIGN FOR BAND STAND IN PUBLIC PARK	222
SMALL TRIANGULAR PARK IN CITY OR TOWN, WITH PLAYGROUND AND WALK (CANAL STREET PARK, NEW YORK)	234
SMALL PARK OF FOUR ACRES FOR CITY OR TOWN (MULBERRY BEND PARK, FIVE POINTS, NEW YORK)	236
TREATMENT OF RAILROAD STATION GROUNDS	243

PART I

THE SELECTION OF HOME GROUNDS

IN attempting to select home grounds suitable for the requirements of a home, one necessarily commences with some more or less vague idea of what the place should be. Generally it is a very vague idea, growing out of partial experience or hearsay, or so-called personal taste. Personal taste is admirable, but only when under the discipline of full knowledge of the subject to which it is applied.

A dozen practical and important considerations may be lost sight of, and one, such as mere beauty of scenery, finally determines the selection of the spot. It is possible, however, to set up some general type or model, the characteristics of which will serve to illustrate the qualifications to be kept in view in all selections of home grounds.

In the first place, all home grounds, especially very small ones, should be comparatively level. Considerable variety of surface and a sky line can be secured by grading and planting the lawn, and the long undulating contours that can be thus secured are more agreeable and restful than the sudden curves of rugged country. Breadth and simplicity of treatment are invited by level or slightly sloping land, and these qualities are indispensable to the restful and pastoral effect which should be sought as essential to agreeable home grounds.

The boundaries of home grounds or village lots count much in the successful treatment of the territories. The shape of the plot should best be oblong or narrow on the street, or square, the latter form being preferable because it admits of more convenient arrangement and subdivision. In the first case, one can lay down as a rule that the lot with seventy-five feet along the street and one hundred feet back can be treated with better effect than if one hundred feet be taken on the street and seventy-five feet back. Of course, it is always possible to meet any kind of difficulty in the shape of a lot, only it is wise not to court difficulties if possible.

More important than the shape of the lot, we shall find, is the character of adjoining property, such as low marshy lands, undrained stables, and all kinds of nuisances to the eye, or ear, or nostrils. On the other hand, one might well wish to be neighbored by a long-tried friend, a noble grove of trees or a beautiful lawn. The neighboring of a sunset sky over lake or ocean, or of a wind-swept field of grass, increases many fold the value of a building lot.

The interior of the place should be considered with the same view to securing simplicity and breadth, for trees will grow large, and houses, in order to carry out the designer's ideas, often take much space. The territory of the home grounds, with the depressions and elevations, must naturally be arranged with room enough for breadth to abide. Tangles, so lovely in wild nature, would be all out of place. They would shut out the light and air, and give a sense of too much confinement; they would bury the house, and neutralize that sense of peaceful buoyancy that comes only with the presence around one of abundance of open space and sky and air. In deference to the necessity for the presence of these essential qualities of breadth and repose, ordinary trees, and even parts of the house, may have to be left out. Perfect scale and proportion are essential on all home grounds, and low, compact trees, like the Japanese maple polymorphum, and the white birch and dogwood, may be associated with low shrubs like Rhodotypus kerrioides and symphoricarpus, and with vines like Rosa setigera or R. wichuriana, giving the effect of a fine miniature lawn picture where breadth and simplicity will still reign in the open stretches of turf. The frame of the picture, the softening of the angles and bare surfaces of the house with vines, may be made delightful on the smallest place by the use of moderate-sized trees and shrubs and herbaceous plants and vines.

To obtain and retain simplicity and breadth, it is well to use a small number of kinds of plants, and to dispose them in borders with slightly curving outlines. Deep bays and recesses of shrubbery, on a small place, may produce an affected and sophisticated, or complicated and confused result, entirely destructive of all suggestions of simplicity.

It is needless to point out that in considering the wants of a country place, small or large, one must first think whether the necessities for comfort and convenience are properly provided for. Room of a suitable character should be secured for vegetable garden, flower garden, stables, chicken yard, and easy turns for wagons, so that the whole establishment can be run smoothly. Trees, rocks, hills, and hollows, on many tracts, group themselves in such a way as to make it evident to the prospective householder that he had better seek elsewhere for what he wants. The spot may be charming and tempting in the beauty it exhibits, but if it does not readily offer on its unchanged conformation the particular features he requires, it would be much better to leave it alone. It is difficult, and often really hazardous, to undertake to change by grading any spot into the sort of place one wants, where at the very outset the natural peculiarities do not suggest the special form of treatment which is sought. There is a forced note about all such work that may mar the quality of the undertaking, be it ever so skillfully carried out. It must be confessed that we are naturally drawn to many lots unsuitable for us to live on, that are in themselves beautiful with hills, rocks, and running or still water. The charms of certain attractive features enthrall us, and we return again and again, in the vain hope that we may be able to think of a way to force its beauty and picturesqueness into the limitations of our home necessities. Finally, perhaps, we yield to the temptation, and buy the fascinating spot. At first it is enough for us to show our friends the many attractive features of knoll and grove and water, but later on, when we start to actually arrange the place with a distinct view to the comforts of daily life, we soon begin to realize the difficulties of our undertaking.

In the first place, where one would naturally seek a building site, it will be found to stand directly on the street and be peculiarly subject to dust and public exposure. The next knoll that has some semblance of suitability for the house will probably be too small, and require, if it is to be used for the purpose of a house site, to have its rounded contours broadened and flattened. But the difficulties will not stop here, for when a course is sought for the road that is to lead to the house, it will be found, perhaps, that some knoll prevents its entrance on the grounds at the best point. When the course of the road is continued farther, contiguous knolls may again make it winding and difficult to traverse with a horse and wagon, and the selection or adoption of some steep grade becomes necessary to reach the house. It is, moreover, a dangerous thing to attempt to radically change the natural contours of any territory, so our way out in building roads and paths and locating houses in such places is not an easy one. In these lands of hills and dales, water will be apt to collect in pockets and threaten us with unhealthy conditions.

Finally, as years go on, we will find that the lawns will not be as enduring under the stress of drought, and the banks more liable to wash, on picturesque hill lots than elsewhere. Indeed, the problem of selecting a home in rugged regions becomes often so hard to solve that an expert may easily make mistakes, for even our greatest architects make them in their most approved city buildings. Common sense, therefore, and a general feeling in favor of economy of effect should prompt us to seek the line of least resistance, and establish our houses and grounds where the conditions readily shape themselves to our hands.

The author does not wish to imply that all lots on picturesque broken ground are objectionable, but simply to explain some of the difficulties that are likely to arise when an attempt is made to create a home on such land. There is little doubt that a level lot is better suited to the general purposes of a home than a hilly one, but, all the same, the reader may come across a property, rugged and broken, which does in the most happy manner fit into his needs for a comfortable and delightful home; only let him be sure he has taken into account all his desires and necessities in this respect, for it is all too easy to forget some of them in the presence of a charming valley or distant view. Yet if he is entirely convinced that he has found such a place, by all means lose no time in securing it and building a worthy house and home.

By selecting a comparatively level open lot where the limitations of the surface are not unduly restricted, the exercise of an intelligent imagination and skill is bound to develop a variety of charm of skyline, lawn contours, and groups of trees and flowers that would at first seem impossible. It is wonderful how much can be done, in this way, by erecting the house on a terrace by means of the earth excavated for the foundation, by lifting the plantations on slightly elevated territories, and by keeping the roads and paths above or below the surrounding ground. Variety is doubtless indispensable to genuine charm, but, on a comparatively level place, it is evident that it may be readily associated with repose and simplicity. On a level lot it is, moreover, easier to shut out disagreeable objects and to develop pleasing vistas and outlooks, and on a square plot the beauty of the exterior as well as the interior effects can be brought out better than on a lot of any other shape. It may take more time to secure the variety and seclusion at all points on a square level lot than on a hilly one, but the trees on a level lot can be placed just where they will produce the best effect as screens and barriers, while the hills, when you commence to plant, are apt to come exactly where you do not want them.

In order to overcome the element of time on the level

lot as much as possible, large rapid-growing trees may be used, at many points, with excellent effect. There is a limit to this transplanting, however, if it be allowable to admit as much in face of various successful removals of very large trees all over the world. Experience has taught men to fix an age and size beyond which it is not wise to move a tree even though previously and lately transplanted and root-pruned. The exact nature of this limit varies almost with every species of tree, and even with different specimens of the same species, where there is a marked difference in vigor of branch growth and multiplicity and freshness of small root fiber. In a general way, it may be said, however, that the large trees of considerable vigor, like maples and elms, may be moved successfully, especially when they have been root-pruned two or three years before, of the sizes of four to five inches in diameter of stem a foot from the ground, and fifteen to twenty feet high, provided they are healthy and full of sap, and not stunted or in any way decadent.

On the other hand, there are trees, like the hickory and pepperidge, that should be set out, after transplanting, of a size not exceeding one or two feet; and the magnolias and oaks, that grow readily only when moved of the small size of six to eight feet. It is necessary to remember, in order to understand somewhat the anomalous results of transplanting trees, that it is not sufficient to make a tree live, but it must grow, to satisfy us. For this reason, one often sees large trees, which have been transplanted ten years, that have scarcely grown a foot; although no one can say that it is not possible to move the largest perfectly healthy trees with a practically unlimited expenditure of time and money. In any case, however, it is a good idea to move trees as large as their nature will readily permit, since the maximum effect is in this way produced in the shortest time, and variety and seclusion obtained in a year or two on an ordinary flat place.

It might seem at first that the presence of woodland on a lot of ground would make it more valuable for establishing home grounds. The shade and natural wildwood effect doubtless constitute a desirable attraction, but even woodland has its decided drawbacks. We may lie on the turf and enjoy the densest shade, for we will readily catch lovely glimpses of the blue sky and flickering sunbeams, but it will be most of the time damp, the grass will be sparse, and instead of young trees and shrubs will spring up brambles and briers. The engrossing presence of the woodland will also tend to destroy that simplicity and breadth that we ought to value so much on home grounds, and surely its sombre monotony of general effect will be apt to mar the place's cheerfulness and variety. It is, moreover, a positive advantage to be able to set out all trees and shrubs freshly in the beginning, for we shall secure thereby more exactly the effect we are seeking, and the time will be comparatively short before the desired effect is obtained. This may seem contrary to the general view of seeking a place which will have trees, houses, rocks, all ready made as it were; but the kind of place on which a home can be best made exhibits originally few incidents of house, trees, or rock, and as few variations of its surface and boundaries as possible. Like a blank page of a book, it will then be ready to receive the full and unrestricted inditement or depiction of the ideas of its owner.

One of the most important practical considerations, and one that should have much weight in the selection of home grounds, is the quality of the soil. A dry sandy loam with a sufficient admixture of clay will give the best results of all kinds for the lawns and gardens of home grounds. Clay suffers from drought, and stony soil is difficult to cultivate in the first place, and in the second place is liable to wash into gullies.

Finally, the author finds himself confronted in the selection of home grounds with the importance of living among people, and of securing the conveniences that tend to make human and kindly everyday existence. Naturally, every one must settle what are his needs, in this respect, in accordance with his own taste, for it is just here that personal convenience asserts itself, and most justly, in perfect freedom. Among such outside conveniences may be mentioned churches, railroad stations, highways, water, street-lighting, and sewerage.

In the present days of many wide-spread advantages, surely pieces of land, both small and large, can be found, by diligent seeking, that will combine with a reasonable amount of beauty and seclusion the essentials of actual comfort and well-being.

THE SELECTION OF THE SITE OF THE HOUSE

IT seems needless to say that the house should be the center and key-point of the entire design of the smallest as well as the largest village lot or country place. A badly located house would evidently destroy the value of the design of the entire place. As the house must remain always the chief consideration of home grounds, its site must be studied from its different aspects, and with regard to its various functions. There are the health considerations, and conveniences of the house itself; the means of entrance and exit, and the exposure and outlook of dining-rooms, kitchens, parlors, and bedrooms. The road and path system of the place, and their connections with stables, drying-ground, and garden, are all important factors in the problem. Distant views will have to be provided for, as well as connections with near vistas, and attractive surrounding lawns, waters, groves, and so forth.

Opportunities will arise, and should be taken advantage of, whereby terraces and level spaces of lawn can be secured for the immediate neighborhood of the house. The contours of the ground should suggest this, and there should be no forced or unnatural scheme of arrangement employed.

SELECTION OF THE SITE OF THE HOUSE

The situation of the front door is the key-note that should largely determine the proper situation of the house in the general harmony of the place. From the front gate the road will lead to the house by a route that will be affected by existing grades and location of trees. The road should seek to conceal itself as it approaches the house, and when the building presents itself in full view, the kitchen, laundry, and the more or less business rooms should not be visible. The main views, such as open terraces, lawns, and specially attractive portions of the place, should lie on the other side of the house, where the disturbing effect of graveled roads will not mar the harmony of the scene, and the guests will feel with unalloyed pleasure the hearty invitation of a hospitable home to enter in and be at rest. Back doors and piazzas will all need consideration in determining the site of the house, as well as the porte-cochère, and subordinate paths and roads leading to various parts of the ground (see pages 4 and 5).

Prevailing breezes in summer are factors in the comfort of home life which should not be overlooked. The house after its many adjustments to satisfy numerous considerations can be generally shifted around so as to secure for the living-room windows the advantage of afternoon shade and cool air. The reader will find on further thought that he can afford to give up advantages such as that which arranges the house parallel with the highway, but he will find the free access of prevailing breezes absolutely essential to comfort in summer life.

Protection from cold winds will also have peculiar importance to any one who spends the winter in the country, or occupies, the year round, a village lot. As

12. HOW TO PLAN THE HOME GROUNDS

the cold winds usually come from a westerly direction, and as the hot afternoon glare of summer comes also from the same direction, it is not difficult to guard

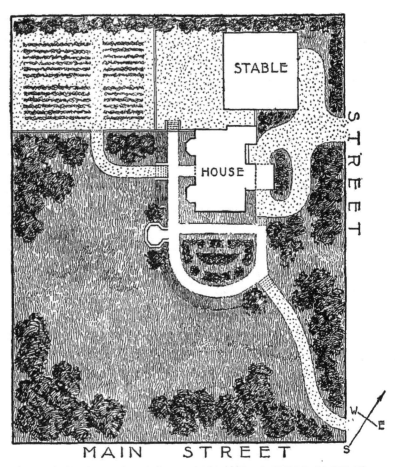

VILLAGE LOT OF ONE-HALF ACRE, SHOWING SUITABLE LOCATION FOR SMALL HOUSE

against both summer and winter discomforts at the same time. The cheerfulness and life of the house depends largely on the morning and noon sunlight being allowed to penetrate the chambers of the dwelling, and to the supreme value of such an adjustment in both sickness and health any doctor will testify.

Having given due weight to requirements of comfort and convenience, we turn once more to the æsthetic side of our problem. To look well and marry itself to the ground, if we may be permitted the use of the phrase, the long way of the house should follow the contours and run parallel with them, and not across them.

LATTICE AND VINE PROTECTION FOR DRYING-GROUND

In view of the necessities of the kitchen, the drying-ground must be kept near the house, and although evergreens and strong bushy shrubs will eventually shut it from view, there will be some time required for their growth to gain sufficient size; consequently, it is a good idea to erect around the drying-ground a lattice of wood or wire, at once, and to plant in front of it rapid growing vines such as honeysuckles, Virginia creepers, and climbing roses, and to front them with deciduous and evergreen shrubs or evergreen trees (see cut above). Inexperienced persons are apt to imagine that trees and shrubs will screen drying-grounds quicker and more completely than is actually the case. Unfortunately, nothing in this climate will screen completely except evergreen conifers such as pines and hemlocks, evergreen shrubs having too low and slow a growth; and even they cannot be always depended on to last many years before decadence and disaster are likely to occur. The climate of America is not, evidently, altogether favorable to evergreens.

So far as the practical features of the house or its arrangements go, there is chance for variety, but within certain well-defined limits. Definite ends must be sought and sure results be obtained.

In considering the practical relations of the site of the house with other features of the place, we should naturally take into account the location of the stable and other outbuildings. There are two ways of looking at the location of the stables. They may be either within fifty or one hundred feet, and it is an old English custom, which is followed with good effect even in these days, to build the house and stables in the same inclosure, so that the roofs are continuous; or the stable may be set hundreds of feet away, screened by large trees and shrubs, on the theory that many things necessarily pertaining to such places would not be agreeable too near the dwelling. At the same time, there is no doubt that a stable can be so well kept as to be almost inoffensive near the house, while, on the other hand, there are many advantages in keeping the stable at a distance, which can be the more readily done in these days of telephones, megaphones, etc. Wherever the stables are, and however much we may endeavor to screen them with trees and shrubs, it will be found a good plan to build wire or wooden lattice-work close to the building, and to cover it with vines so as to embower it more completely than the trees or shrubs will be likely to do for many years.

In any case, care should be taken to arrange the stables so that they seem either a part of the domain of the house, or appear evidently detached and associated with something else at a considerable distance. It would be unfortunate to locate them in a sort of semi-detached way, isolated, with no apparent connection with the house, which may mean that they are neither sufficiently near nor sufficiently far. There is an unrelated, detached way of arranging the features of the grounds that will, if used, give the stables a lonesome and inconvenient appearance, and distinctly mar the general appearance of the place, and it is against this method, or lack of method, that the reader should be specially warned.

But when we come to contriving various adjustments looking to the development of the beauty of the place, variations of effect may be managed in a dozen ways. By shifting the house a little, fine views heretofore hidden may be opened from the piazza or the porch. Vistas and glimpses of scenery may be made to suddenly appear, or only creep gradually, as the hours of the day and the seasons change, into view of the windows at which the family spend most of their time. It would be enough to repay much thought and contrivance if only one noble tree or fine massive rock were, by some special adjustment of the house site, brought into view of the dining-room or sitting-room. The same line of study will lead to the contrivance of open spaces of level lawn around the house, where an expanse of turf, raised on a terrace, will lend dignity and distinction to the building. The road, inspired by the same desire to develop in every way the possibilities of beauty and convenience inherent in the site, may be lead by tree-masked ways to a point where the house will gradually glide into view, bit by bit, till its full effect stands before us.

In the case of a small village plot this complete masking of roadway and path is seldom advisable, but generally one or more large trees can be so placed in relation to the house that the breadth of its general surface, broken up by masses of foliage, will serve to reveal only small areas, here and there, of the roof and sides.

All these changes of the original earth surface may be so managed that, when the house is once located, the remainder of the ground seems comparatively unchanged. At intervals a road or path will peep out in an unobtrusive way, and open spaces will appear only as natural-looking glades or lawns, or as a suitable resting place for the home itself.

The turf and trees and shrubs will seem to be a natural arrangement of features that have apparently, by accident, fitted themselves to the needs of the house. This kind of work is not expensive, and all the more dignified and refined because it adheres as closely as possible to the original peculiarities of the ground. A few trees or shrubs are set here and there, or in borders, simply to help out the natural suggestions of the place, and all too obvious effects are obscured as much as possible by various devices of arrangement.

On a village lot this obviousness is more difficult to manage than on a larger place, and obviousness in excess is objectionable, because all parts of the place should blend in gradual and perfect harmony. It is an advantage to the general effect of a small building lot to set the house well back from the street, and to mass the front and side lawns so as to give breadth and simplicity and as much depth as possible. To secure this effect in the best way, self-restraint must be exercised in setting out trees. It may seem a little strange to the reader to be warned against setting out trees and shrubs, but it is easy to conceive a small house lot where the proper simplicity and dignity of the place require the plant adornment to be limited to turf on the lawn, and vines on the fence and house.

In the same way these small places should be graded level, and not scooped out in valleys in a forced and unnatural and often undignified fashion. Simplicity and dignity should be the key-note of all landscape gardening, and from the nature of the plants, variety will be sure to follow with happy effect the presence of the few trees and shrubs and vines to which we may be obliged to limit ourselves.

These preliminary principles being laid down, whoever adheres to them, when he assumes the responsibility of selecting a house site, will find that the comforts and delights of living will be reasonably provided for. It only remains for him to successfully pursue his own individual taste in the selection of trees and flowers, and the readjustment, as time goes on, of the roads, paths, and lawns to any altered conditions that may arise, provided he will never forget to be controlled by these simple and common-sense basic ideas.

ROADS AND PATHS

WE have already found that turf may be much more attractive than roads or paths, even though the latter are planned for æsthetic reasons as well as for the sake of convenience. The cost of mistakes in ordering such matters too exclusively or unintelligently on the artistic side, is some of the actual convenience of the place; the cost of mistakes, when the beauty and artistic effect of the place alone are considered unintelligently, is the constant irritation, and consequent loss of comfort, of things not rounded out and made suitable for the definite end for which they ought to be made.

We shall see, as we consider further the different features of home grounds, how often the question arises as to the best way to reconcile apparently conflicting claims of beauty and utility. Each should have due consideration, each should be studied in the light of what we believe to be sensible needs and conveniences and genuine artistic simplicity and breadth, and these need not be allowed to conflict with each other if they are intelligently managed. Thus the temptation comes to multiply the number and extent of our roads for the sake of convenience, with the sad result of mighty small benefit in the way of convenience and much permanent injury to the beauty of the surface where unattractive roads take the place of attractive turf or trees and shrubs.

In like manner it is frequently most natural and most convenient to follow a comparatively straight or slightly curved line to the house, except on a very small lawn where the necessary economy of limited space may restrict us to a perfectly straight line; and here again, no temptation of the beauty of curves that deviate widely from the straight course should be allowed to divert the more sensible because more direct course. For the same practical reason the apparent æsthetic value of winding or twisting roads, in spite of their frequent beauty of line, is doubtful, when we consider that they are likely to be inconvenient and difficult for the passage of horses and carriages. With paths it is different; a path may gain by winding and twisting, provided a reason for the curve is evident in the shape of a tree or rock, or some particular view of building that it is desirable to reach by a sudden deviation of the line of travel.

It is easy to conceive of many small places where a straight road entering the center of the grounds is the only thing advisable (see page 28); with large estates this is not so. It is well to keep one's mind free from prejudice in favor of formal styles of treatment, depending entirely on the suggestions of the ground and not on a prior theory. An excellent way to design entrances, where the necessity of driving in either way is of equal importance, is to go in at two points, as on page 22, thus creating an enclosed territory or ante-park that offers itself for planting, as a foretaste, as it were, of pleasure, before the place itself has been reached, and helping to make, by the employment of large masking plantations

20 HOW TO PLAN THE HOME GROUNDS

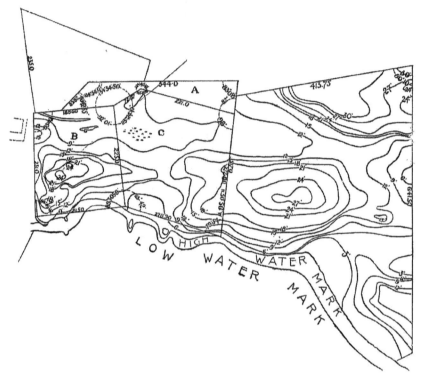

CONTOUR MAP OF COUNTRY PLACE OF EIGHT ACRES, WITH STEEP CONTOURS

of shade trees and shrubs, one of the most agreeable approaches to a country place that can be imagined. The entrance in this way gains dignity, and a rural and park-like character that has none of the pretentiousness of some cut-stone and elaborate iron gate designs.

In studying the different elements that contribute to the efficient and harmonious development of all country places, we find that three notable features present themselves for our consideration, namely, the house and its outbuildings, the area of the lawn, and the area of the trees and shrubs. In the interest of proportion and

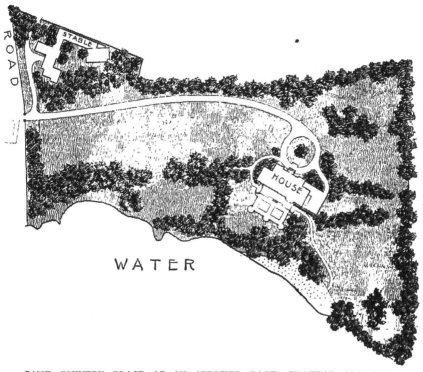

SAME COUNTRY PLACE AS ON OPPOSITE PAGE, SHOWING ARRANGEMENT OF PLANTATIONS, ROADS, AND PATHS

harmony, the space alloted to these features first needs study.

These three primary and all-important elements of a place having arranged themselves in a definite and rational fashion, the consideration of the roads and paths comes next. The general idea of them is convenience and beauty, which is synonymous with simplicity and

SECTION OF ABOVE PLAN TAKEN ON LINE ACROSS THE CENTER OF CIRCLE, HOUSE, AND TERRACE

directness. There is always a particular curve that suits a road in a special region of the grounds both for beauty and simple directness, and there is a special width needed for the road that is at once convenient and looks in due proportion to the general appearance of the place.

By commencing at the entrance, and marking out a way to the front door, we will discover some of the limitations that will control the course of the roads and paths. The entrance will have its relations to the highway or nearest railroad station, and its position will be largely determined by the nature of the ground, whether

DOUBLE ENTRANCE, WITH ANTE-PARK

hilly or otherwise, on the boundary of the property. It is usually better to enter a place on one corner or the other, and reach the house by easy and slightly winding curves, or almost straight lines (see page 21).

When one assumes this attitude of mind toward all landscape problems, it is wonderful how many ways will present themselves for accomplishing a desired result, and how beautiful and simple and convenient will be the plan finally adopted if we will yield ourselves to the undivided teachings and suggestions of the environment itself.

To enter the place agreeably and safely, and prevent

as much as possible all danger of collision with teams coming the opposite way, it is wise to carry the road directly into the grounds, at right angles to the highway, twenty-five feet to one hundred feet, in accordance with the size of the place (see page 25). When the road turns, it should skirt around, and not across the lawns more than is necessary to make a reasonably direct course to the house. If it runs into and through the mass of shrubbery somewhat, it will tend to secure for the road a partial concealment, which we have seen is also desir-

CARRIAGE TURN WITH GRASS PLOT, SHOWING COURSE PARALLEL WITH HOUSE

able. Wherever some natural obstruction, such as house, tree, or rock, does not force the road into a sudden curve, the aim should be to bend it into a long line, as nearly straight as the circumstances will permit. Reverse curves or snake-like twists, as we have seen, are apt to mar the simplicity and dignity of the design, and, above all, to confuse the horse unless he is driven by a particularly skillful driver (see page 24).

As we approach the house, we will find that any curve in our road must be entirely straightened, and a line taken parallel with the house throughout at least the portion of the front it passes (see page 24). This kind of approach possesses more dignity and presents the house to better advantage than the one that leads up to the house and directly away again, by means of a circle or narrow turn. If circumstances will allow—and it is wonderful how much circumstances can be made to allow to the intelligent effort of careful study—it is a good idea to design all curves of roads on different lines of the ellipse, and to avoid arcs of circles, because they are difficult, in most cases, to employ in a practical way.

Whether the road shall enter in front of the house and come out again at the same point as it entered, or

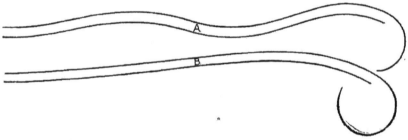

A—BAD LINE FOR A ROAD; B—GOOD LINE FOR A ROAD

whether it shall leave the place at another and more distant place, are questions which the size of the grounds and other considerations will govern. It is safe to say that, all other things being equal, the best road is the one that takes the shortest line. Sometimes a simple widening of the road in front of the house simplifies matters and gives sufficient room for turning (see page 26). It is impossible and unwise to attempt to say how every special problem of road arrangement should be treated, for we cannot assume to know all the circumstances that will control the result. It is wise, however, to say that the width of the road should be minimized as much as possible, and that its course should lead directly toward a

ROADS AND PATHS 25

definite goal, and move with long easy curves; but after that is said, we should pause and await the presentment of all the factors that enter into the problem before deciding upon the line to be taken. It is a wise man who uses rules when he needs them, and is not slavishly controlled by them.

Plan on page 28 shows a village lot properly laid out with straight lines, with not a curve among them, and yet we have recommended the use of forms of the ellipse,

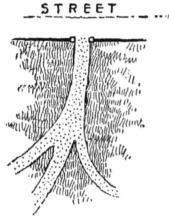

BRANCHING ROADS IN LARGE PLACE, ENTERING AT RIGHT ANGLES TO HIGHWAY

so it is evident that circumstances must always control largely in any scheme or system of roads and paths.

The value of the contour map becomes evident when we undertake to arrange the road and walk system of the smallest village lot. The lines as seen in plan on page 20 represent circuit lines or curves of the surface of the ground that exist at a uniform level, or, in more scientific phrase, intersections of the surface of the ground with a series of horizontal planes at equal distances apart. In the diagram of the map shown on page

20 the contour lines or intersections are three feet apart. In working out the contours of the ground of a small place, we may seem to be taking trouble that is hardly necessary. But it is in reality a sure way of securing the best and most certain results. When we find in this way that a road cannot be carried a hundred feet on a grade less than ten per cent. without filling in earth, we will be in better shape to solve the road problem than we were before we obtained the contour map.

If we work out from the contour map parallel cross sections of the proposed roads at different points, and plot these sections to a scale, in their true relative posi-

OPEN CARRIAGE TURN ON MEDIUM-SIZED PLACE

tions, or referred to the same level or datum line, it will be easy to locate the axis of the roads and estimate the quantities of excavation and embankment. In this way it is often possible to learn that the fill and excavation can be made exactly to balance each other, a fact it would not be possible to learn in any other way.

By taking such precautions to secure an intelligent plan and estimate of the special details of a road system, the owner of a village lot can always learn what the entire undertaking will cost, and pursue his work in a sensible, practical manner. Village lots have been, in the author's experience, more often laid out with a lack of an intelligent plan and estimate than ordinary country places of a number of acres.

It is difficult to decide upon a standard of road grades for ordinary places. So much depends on the kind of vehicle for which the road is chiefly intended, on the character of road covering used for the surface, and on the condition in which the surface is maintained. The grade should not be so great as to require the application of the brakes to the wheels in descending, or to pre-

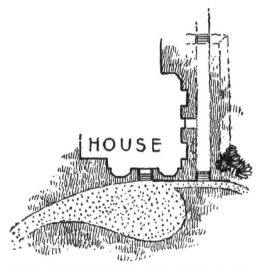
OPEN CARRIAGE TURN FOR SMALL PLACE

vent an ordinary vehicle from carrying a load of passengers with horses ascending the hill at a trot. Generally speaking, the grade should be somewhat less than the angle of repose, or that angle upon which the vehicle in a state of rest would not be set in motion by its own weight, but would, on slight motion being imparted to it, descend with slow uniform velocity.

In practice, the steepest grade that should be allowed on macadamized or telford roads such as are generally used on country places, is about one foot in twenty, or five per cent. It is entirely practicable to use much steeper grades, but it will, probably, be done only with the assistance of a brake going down hill and at the expense of a slow walk up hill. The grades one foot in thirty feet, or one foot in thirty-four feet, or about three per cent., are most desirable, because then the speed ascending need never be slower than a trot, and descending will never require the application of the brake.

In constructing the roads of a place, the excavation and embankments at once call for attention, for on them depends the preparation of the roadbed whereon is to rest the stone structure or metal of the road. If we are not so fortunate as to have our cuts and fills balance each other, we shall be obliged to cart our surplus material to the nearest point we can find, and obtain or make convenient places from which to cart additional soil. In making these embankments it should be remembered that different kinds of earth do not fill the same space in artificial embankments that they did in their natural bed. The increase in volume of freshly dug earth often varies twenty per cent. among the different kinds; but, curious to relate, when formed into embankments, it shrinks to less than its bulk in the natural bed.

In excavating and moving earth, it is first loosened with picks, shovels, or plows—the plow is very useful—and then shoveled into carts or barrows and taken away. For short haulage, say ninety to one hundred feet, the ordinary road scraper, holding about one-tenth of a cubic yard, will be found useful. It is not profitable to work the road scraper over ground that is steeper than one foot in five feet, or twenty per cent. For dis-

VILLAGE LOT, EIGHT-TENTHS OF AN ACRE, WITH OUTBUILDINGS
AND STRAIGHT WALK SYSTEM

tance exceeding the sphere of scrapers, earth is generally conveyed in wheelbarrows. The limit, when one-horse carts should replace barrows, will seldom exceed two hundred and fifty feet to three hundred feet for all the various kinds of earth. Beyond a certain distance, determined largely by the character of the road and its grade, two-horse wagons should take the place of carts.

Having finally adjusted all questions of grade, excavations, and embankments with the assistance of the contour map and experimental lines on the ground, the course of the road adopted should be carefully plotted on the map, together with cross sections that will show the cuttings and fillings as well as the natural surface of the ground. Specifications of the different kinds of work, and drawings of drainage and pipe lines, culverts, etc., should also be prepared. The center of the road can be located on the ground that will correspond with different points on the map. *Cut* and *fill* should be marked on the stakes to indicate that the natural surface should be cut down or filled up at these points in order to secure the proper grade of the road. Stakes showing the proper width of the road should also be set. All this work of planning and preparation may seem unnecessary on a small place of less than an acre, but it will be found to fully repay the trouble, for only by making proper plans and specifications can you be sure beforehand of good construction.

The temptation to construct steep banks, to save expense, often leads to much trouble in droughts and heavy rains. In dry weather the grass is especially liable to turn brown and die on such slopes, and rains are generally apt to gully any slopes that are steeper than forty-five degrees, or one foot on the perpendicular to one on the horizontal. Ordinarily the one-to-one slope is steeper than is advisable, and one and a half to one is considered as steep a grade as should be given banks. The sliding of soil on a steep grade may be largely prevented by covering it with a foot of rich mold or loam, and sodding or seeding with grass will also tend to make it still more firm. Sodding for slopes is greatly to be preferred to seeding with grass seed. A steep bank may be strengthened by a low stone wall at its base that will serve to restrain it from slipping.

If the soil of the bank should be infested with springs which would be liable to impair the firmness of the slope, they should, if practicable, be tapped at their source, and the water conveyed by stone or tile drains to the gutters along the road. These gutters or drains should also extend along the top of the embankment, to prevent the water from coming over and down the slope. The cut of a road can usually be made to equal the *fill* by locating it in just the right place, but where it is very steep, say forty-five degrees, or one to one, rough dry retaining walls should be made at the bottom of the embankment.

Drains will often be required at a little distance above the cut of the hillside road. If the road should run through a marsh or swamp resting on a firm bottom, the soft material should be dug out, provided it does not exceed three or four feet in depth, and the road built directly on the hard soil. Deep open ditches should be dug on either side of the road, and sometimes, when the neighboring wet land is especially deep, cross drains under the road at frequent intervals become necessary. The drains may be of stone or tile. When the marsh is very deep, it is sometimes necessary to build under the entire roadbed a heavy mass made of bundles of twigs or branches, each bunch ten inches in diameter and twenty feet long. These bundles are laid one across the other, layer after layer, until the top layer lies transverse to the direction of the road. Usually the ditches at the side of the road are made open.

The waterways of culverts should be large enough to take the greatest volume of water they will in all probability be required to carry off. Eighteen inches, or, if their shape is circular, twenty inches in diameter will usually suffice for this purpose. Small culverts are often made of slabs or plank, but such methods are shiftless and are not to be commended.

The drainage of the surface of the road on ordinary places is generally done with paving blocks of granite, trap, or asphalt composition about the size of an ordinary brick, set on edge, except that they are made considerably thicker in order to bed them properly. They are laid in sand, and thoroughly rammed down on the foundation or metal of the road, which should be in all cases carried across under the entire gutters so as to prevent all chances of the gutters settling. These gutters are usually made fourteen inches, eighteen inches, and two feet wide, according to the size of the road. A slightly concave surface should always characterize a gutter, but the mistake is often made of hollowing them out too deeply.

It is easy to see that the gutters would be of little use unless they were connected with a complete drainage system, through road basins located twenty-five feet to three hundred feet apart, into drain pipes that will lead the water to some general sewer or waterway. But it is naturally asked by many who have moderate-sized country places or village lawns. Why can we not dispense with these gutters? They are expensive, and are, to say the least against them, not attractive looking. In regard to the drainage, the pipes and road basins, the author can suggest no way of dispensing with them; the water must have a way to flow off. But many of our readers may not be aware that there is such a device as a sod gutter, which, if properly constructed and connected with the drainage system, will perform its duty quite as well as the ordinary stone gutter, and be at the same time more attractive and economical. The trouble with sod gutters generally is the danger of gullies forming in them by sudden floods of heavy rain. This can be pre-

CROSS SECTION OF ROAD WITH SOD GUTTERS

vented by attending to thorough ramming of small, thick, strong sods on a gutter that has only dish enough to collect the water, and, in practice, it will be found that a very slight concavity will accomplish a satisfactory result; but this all presupposes that the drainage system established is perfected with field basins and road basins near enough together, and a sufficient number of other outlets, to properly care for all the water that flows from the surface of the road.

It is often a good idea to lay two-and-a-half-inch agricultural round tile without collars a foot under the low part of the gutter, to lead the water that flows from the side slopes away from the sod gutters and the roadbed to the basins and pipe drainage system. It is wonderful to find how perfectly a system of drainage of this kind protects the road, and how attractive the borders appear when compared with the effect of the ordinary stone gutter. It may be said, and possibly conceded, that there are slopes on roads so steep that only a stone gutter will accomplish satisfactory drainage, but the reader may be sure that there are very few grades actually found on roads whereon the sod-gutter system cannot be used satisfactorily if properly and skillfully constructed (see page 34).

So far as the width of roads goes in large cities, it is well to be liberal. One hundred feet is not excessive for streets that are leading arteries. On ordinary country roads, however, a width of sixteen to seventeen feet is enough for the actual roadbed, exclusive of gutters and sidewalk—two teams can readily pass on this width—and on village lots, roads not exceeding thirteen feet or even twelve feet wide can be made to suffice, in view of the fact that most vehicles have an extent of five and a half feet to six feet from the outside of one hub to the outside of the other. The larger the place, the wider should be the roads up to the point where it is felt the harmony and picturesqueness of the place will be injured by the obtrusive size of these features, which must, in the nature of things, detract from the scenery, and should be only tolerated because they are necessary.

There is a difference of opinion among experts in regard to the best method of crowning a road. By one method a cross section of the road is made to exhibit a convex curve or a semi-ellipse, while by another, bearing the weight of testimony in its favor, two equal planes of the surface slope gently to the side gutters, and meet in the middle by a short connecting flat curve. This method will lead carriages to easily drive on the sides rather than on the middle, where collision is apt to occur, and will avoid the tendency to overturn or slide sideways near the gutters. This construction may be described approximately as consisting of a surface which is defined by two straight cross lines connected by a flat arc of a circle on a road which in ordinary cases should be sixteen feet wide. The inclination of the surface from the end of this arc to the gutter, for rough earth roads, should be one foot in twenty feet, and one foot in thirty feet for ordinary gravel or broken-stone roads.

It is evident that ordinary earth roads without a broken-stone foundation can never be satisfactory at all seasons of the year, but where we are obliged, on account of their cheapness, to use them, we should remember that attention to certain features of construction is always important. In the interest of good drainage, the ditches along the sides of the road should always be kept open, and sufficient slope given to allow the water to run freely. Hollows and ruts should be filled up as fast as they are formed, and the customary rounding up of the surface familiar in country districts avoided, and, as already directed, a slope made from the center to the sides that will not exceed one foot in twenty feet. It should be needless to explain that earth roads above all others should not be steep, on account of their imperfect surface, but, unfortunately, it is the earth road that generally presents the heaviest grades.

We find another hurtful practice common in the treatment of earth as well as stone-bottomed roads, and that is the scraping up of waste material for use on the traveled surface of the highway. It is important, in every case of repair where extra material is needed, to secure fresh gravel or broken stone that has been properly screened or otherwise prepared.

Graveled roads have been found to be well suited for country places, although they are not generally as reliable at all seasons as those made of broken stone and screenings. The clay that is necessarily left in the gravel for the purpose of binding it will always tend, in certain seasons, under the wheels of carriages, to grind into mud. It is difficult, often, to secure just the right kind of gravel for road-making, for clean material consisting of round pebbles will not pack, because a certain admixture of clay and some angularity of the stones are necessary to secure a proper bond. It has been found that seaside and riverside gravels are too clean, and ordinary pit gravel is too dirty or clayey.

By using two wire screens of the proper size we can secure with many pit products what has been found to be the best kind of gravel; namely, a moderately clean article containing no stone less than half an inch in diameter and none more than an inch and a half. This is done by placing one screen above the other, the lower one allowing everything smaller than half an inch to pass through and the upper one retaining everything over an inch and a half. It is no objection, and rather an advantage to the screened gravel, that considerable clay is sure to adhere to the stones, for on the presence of this clay depends the capacity of the material to properly bind together. Much skill is required to secure just the right admixture of this clay in order to prevent as much as possible both mud and dust.

On ordinary soil, an excavation to the depth of ten or twelve inches will suffice to make a gravel road, and you may, in dry ground, where the travel is light, construct a good road with the entire twelve inches of gravel; four inches of it being ordinary pit gravel, and then two four-inch layers of double-screened gravel, each in turn well rolled. A better gravel road than this is used in many private and public places, where the bottom is constructed of five or six inches of broken stone of equal size, placed regularly by hand and bedded and rolled, with the addition on top of four or five inches of properly screened gravel, also well rolled. The presence of this stone at the bottom makes better drainage for the road than if it all consisted of gravel. In spring, moreover, when the frost is coming out of the ground, the clay is apt to work to the surface and create more mud where the stone foundation is lacking than where it is not.

It may seem to require a great deal of work to build a road of this kind, but, when done, it is without question superior to the ordinary gravel road, which is made by heaping unscreened gravel three to four inches deep over a width of eight or nine feet, with perhaps six or ten inches under the wheels. There is sure to be plenty of mud and ruts on such roads, and, consequently, continual need of repair. To make sure of a thoroughly solid foundation of stone, care should be taken to keep the fragments of nearly equal size, and to chink in between them slivers or spalls.

Mention has been made on several occasions of the importance of rolling a road at various stages of its construction, and it needs explaining, in connection with the rolling, that at first water should be sprinkled in comparatively small quantities on the surface, increasing the amount gradually, until the finishing strokes are given in floating water rising in a small wave before the roller.

The roller should be started at the side of the road, thus pushing, as the roller moves back and forth, all the surface material more or less toward the center. Skillful and persistent rolling, sometimes for weeks together, constitutes a large part of the secret of the construction of a good road, whether it be made of gravel or broken stone and screenings.

A pernicious way of building roads is often adopted by spreading a thick layer of clayey loam on top of the rubble or stone foundation, to make what is termed a cushion. The difficulty with this kind of work is that clay is liable to work in large quantities to the actual surface of the road, and allow holes to form by settlement. A reason for the popularity of this so-called cushion may be found in the fact that it is easy, in this way, to fill with earth, cheaply and readily, the interstices of the stones which should have been chinked up with slivers and chips.

A good rule has been laid down, which may be modified by circumstances, as all rules must be, whereby a road shall have a depth of foundation of rubble or telford of not less than one-half and not more than two-thirds of the entire thickness of the road covering. In this way a road excavated nine or ten inches deep may have a foundation of large stones six inches thick, while seven or eight inches would not be too heavy for a road excavated twelve inches deep. The difficulty with a road, the foundation of which is not sufficiently bedded and chinked with chips and spalls, is that a heavily loaded vehicle is liable to tilt up one of these imperfectly bedded stones, and thereby render necessary the reconstruction of the entire driveway at that point. The foundation stone should be of as nearly equal size as possible, vary-

40 *HOW TO PLAN THE HOME GROUNDS*

ing, according to the depth of the road, from three to six inches deep and eight to twelve inches long.

Shell roads are valued for sandy soil, and do well for light traffic, but they need constant repair and renewal.

The best treatment of footpaths will be found, in the

VILLAGE LOT, ONE-HALF OF AN ACRE, WITH STRAIGHT WALKS

main, nearly the same as that we have been advocating for roads, although they need never be as thick and solid. Gravel and broken stone are, it is generally found, sufficiently dry for walking, and the tone of color of such walks is more agreeable to the eye than that of asphalt or brick. On village sidewalks, it is needless to say, the asphalt and bricks are to be preferred.

It is a good idea to lay a row of sods along path and road borders, as it is not practicable to get firm edges by means of grass seed. So far as the paths are concerned on a country place, they should be run entirely independent of the grade of the roads, although their course may be near them. Diversity of surface is generally agreeable, and it is entirely unnecessary to seek parallelism, or even similarity of level and course, in the construction of lawns and roads and paths. The only thing to avoid in design is the close approach of roads and paths, whereby the ground is forced to assume the shape of very steep banks.

The mistake is often made of allowing the road to continue a long time in a bad condition, which always involves more actual expense when the work is finally taken up. Day-by-day maintenance, it is evident, will either prevent disintegration of the road surface, or will, by incessant work, prevent the extension of injuries that would grow to serious dimensions in the course of time. The prevention of wear and tear is, we will find, attained to a large degree by sprinkling with water the smooth surface of roads, thereby laying the dust, and helping to hold firm the cementing or binding quality of the road. The sprinkling should be given lightly, so as not to create mud and ruts. The best sprinkling carts are those that can be made to distribute water evenly and lightly, for it is easy enough to make a cart that will sprinkle heavily.

Good maintenance, also, naturally requires continual attention to open ditches, gutters, roads, walk basins, and drains of all kinds, and, above all, attention should be given, day by day, to the minute repairs of small patches that are intended to fill up all depressions and ruts as soon as they present themselves. In this way, the road can be kept undiminished in thickness for an indefinite period. The repairs should be so managed as to expose the road, whether gravel or broken stone, slightly to view, without laying it bare or removing the binding material from around the stones at the surface. This may be done by men suitably provided with hoes, stiff brooms set at right angles to the handles, shovels, and wheel-

FORMAL ENTRANCE OF LARGE PLACE

barrows. Machine scrapers and brooms of various kinds drawn by horses have to be used with great care, to prevent injury to the roads by loosening the top stones. A certain amount of detritus, or binding material, should be left on the road, and frequent rolling in the spring of the year will help to make solid the surface that has worn loose during the wear and tear of the frost and travel of winter. To maintain a road properly on even the smallest place, the rolling should be carried on systematically at regular periods all through the year, but more especially in spring, just after the frost has disappeared from the ground.

The importance of systematic maintenance seems to be understood tolerably well in the case of lawns, but with roads it appears to be different. To carry out the work of maintaining roads properly, a laborer should be given charge of an allotted length of road, on the *block* system, for the proper repair and cleanliness of which he should be held responsible. His duty should consist in keeping the road always scraped clean, and free from mud, and in filling in ruts or hollows, the moment they appear, with broken stone or gravel brought from a storage place near by.

On all repairs of roads, as much of the old, but not

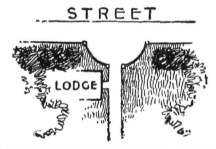

TREATMENT OF ENTRANCE GATE AND LODGE

waste, material should be used as possible, the object being to unite in bond the old material with the new, so that the patch will be as little unlike the unrepaired portion as possible. Experience will soon enable the laborer to judge when the old material is too much worn to use again. In making these repairs a roller is very useful, although excellent work can be done on broken-stone roads with a rammer weighing twelve to twenty pounds. The wheel tracks should be promptly obliterated and filled in, and the new stone, that has just been applied wherever it is needed, should be covered with fine material and rammed and rolled in with the help of a light sprinkling of water. Moderate sprinkling in dry weather is especially necessary to keep the road in order, as the drought bakes the surface and makes it wear away faster, and sometimes loosens the small stones.

To construct a good road requires skill and knowledge of the comparative value of different materials, but to maintain it properly through many years renders necessary an exertion of skill combined with intelligent, persistent diligence that is not, in this country, as often exhibited as we should like to see it.

LAWNS

IN the eyes of most people who have had no experience of their own in these matters, an open stretch of turf is a comparatively blank space that offers itself more or less strenuously for occupation, as is exemplified by the frequent efforts of citizens to employ park meadows for public shows, race courses, and menageries. It would really seem as if an open lawn, with all its beautiful expanse of turf, emerald-hued and cloud-shadowed, would impress wholesome-minded human beings as something more than a neglected opportunity for some building, glass house, or flower garden; but everything depends on the standpoint from which one looks at a question. If one is thinking of deer paddocks or flower gardens, a vacant piece of grass suggests only the opportunity for promoting the favored object. If the eyes of ordinary observers can be brought to dwell on grass space as a strictly beautiful object which they ought to be able to appreciate, open meadows will soon become for them also supremely valuable possessions.

After falling for a while under the spell of their gentle and quiet pastoral charm, one will feel that the very heart of the landscape picture lies within the tender green space, the delicate refined quality of which has, just because it is refined, escaped observation, and that is why the author places the lawn next in importance to the house. Indeed, the arrangement of trees and shrubs performs its special office when it serves to develop and heighten the attractions of the lawn. When one thinks of the true function of the lawn, the vision arises of a masterly painted canvas whereon are depicted moving cloud shadows, waving grass, rich patches of dark and light green, studded with the starry radiance of the humble flora of the grass, and the hundred incidents of blazing or subdued color and form that appear on the surface of an open meadow.

The outline and variegated boundary and frame of all this loveliness is the trees and shrubs, the varied masses of which serve to emphasize and reveal the most evasive charms of the territory. The concealment of the roads and walks by various tree and shrub devices seems the more necessary the longer we contemplate the special beauty of the lawn, and feel the less fine quality inherent in roads, which have value mainly because roads are necessary for the proper enjoyment of the place.

It is around and about the house that the open expanse of lawn performs its most important function (see page 47), lending such dignity and effective presence to the building that, unless some special peculiarity of the ground prevents, it is the lawn that secures the most admirable and rational setting for the house. For a similarly deduced reason, on comparatively flat territory, the erection of the house on a lawn rising partly in terrace shape gives the home a distinction that adds greatly to its attractiveness. The use of the terrace effect needs handling with extreme skill and care. The proportions, shape, and construction of terraces, formal or otherwise, on rugged hillsides will have to be managed with good judgment and adroitness. It is so easy to do violence to the most attractive suggestions of the ground itself.

VILLAGE LOT, ONE-THIRD OF AN ACRE, WITH OPEN LAWN, OVERLOOKING FINE SOUTHWEST VIEW

Many devices of curving banks and irregular low masses of shrubs may be so employed as to produce the effect of a terrace, without disturbing the essential character of the region. Whereas, by setting flower beds and specimen plants here and there, just because

48   *HOW TO PLAN THE HOME GROUNDS*

one likes flowers or rare plants, is pretty surely apt to work to the detriment of the effect of the general expanse of the lawn. Haphazard and unformulated planting, whether thick or slight, is sure to work against

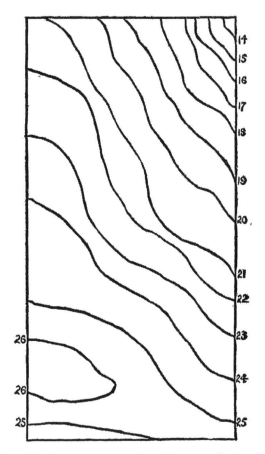

CONTOUR MAP OF SAME VILLAGE LOT, ONE-THIRD OF AN ACRE

simplicity and beauty. Of course, fine single specimens of trees and shrubs may find points suitable to receive them near the house, and so may flowers, but the choice of their place must be indicated by the nature of the ground in that special neighborhood. In a general way, they had best be kept some distance away, or located near some mass of foliage, so as not to mar the open character of the space which is really the heart of the home grounds.

As the expanse of turf leaves the house, the spread and breadth of the lawn should, if possible, increase. It is wonderful how, on a very small place, the landscape idea being kept strictly in view, the most picturesque effect can be obtained by a few skillfully arranged shrubs and trees and a skillfully graded lawn.

By undulating the surface of the lawn toward the trees and shrubs on either side, and keeping the planting territory elevated, and grouping the bordering foliage into points and bays of green, pictures can be created upon a limited space that would delight the eye of a painter.

The grading of the lawn is a delicate operation. It is not difficult to grade a lot level, or sufficiently level for good effect, the production of a mathematical level being practically impossible, but where the difficulty commences is when nice modelling of the surface of a territory is directly suggested by the natural scooped-out character of the ground. The thing to do is to catch the idea of the slope and swing of the ground, and instead of trying to make it more level, to rather accentuate the dip of the general curve of the surface exactly in the trend and direction suggested by the surrounding land. The general curve should be made, of course, continuous, and all little hills and hollows smoothed out into one suave and well-modeled valley.

Low, marshy ground may be often drained dry instead of being raised higher with earth filling, and rock masses draped with vines and planted with low shrubs instead of being blasted away. In this manner, the lawn and other parts of the place may be often brought into harmonious relations without sacrificing a bit of the original charm of the surface of the territory.

Perhaps it may not be amiss at this time to further explain the position taken above, by reducing the theory of primal arrangement to the following simple terms, provided they are made subject to considerable modifications that circumstances may render necessary. Given an open lawn, and a house, and a plantation of trees and shrubs on the boundaries, and the place will be essentially complete. The roads and paths simply serve to link these features conveniently together, and being no integral part of the artistic or pictorial design of the place, should therefore be screened and kept out of sight as much as possible, and run where they can be on one side and go through the bordering plantations.

After all, there should be allowance made for the peculiarities of different places, which have to be met in different ways, but when we come to the actual construction of the lawn, the preparation of the ground and its enrichment and seeeding, the practical operations must be carried on according to certain tolerably well-fixed rules.

Depth of cultivation, it need hardly be said, will be found to be almost indispensable to the creation of good sod, and every pound of superphosphate of lime, bone-dust, or well-rotted manure—amounting to, say, a maximum of a ton of phosphate of lime or bone-dust, and fifty tons of manure to the acre—will make itself evident in good results. The exact amount that it will be found profitable to apply will naturally depend on the kind of soil intended for its reception. It may seem unnecessary that this advice concerning the liberal application of soil and manure, which has been familiar to cultivators almost since man tilled the soil, should be repeated so strenuously, but in spite of its familiarity the subject needs, in all seriousness, dwelling on again and again, and year by year, and probably will require it as long as the attempt to make lawns continues to be made.

The application of the fertilizers should be made, and the ground plowed and harrowed or raked, just before the grass seed is sown, unless the additional precaution is taken, especially on old sod ground, to give the land a coat of manure a year before, and secure its full effect, and also a desirable condition of tilth and destruction of old sod by the cultivation of a potato or corn crop. When the ground is tilled and manured, the choice of the grass seed to be used should be made with the greatest care.

Notoriously, grass seed is apt to be full of weeds and chaff, and the very best quality is therefore cheap at what may even seem an exorbitant price. It is better to buy the cleanest and best red top or blue grass or bent grass, and either mix or use them alone. They are strong-growing grasses, and will very likely outgrow, in a short time, any more attractive but weaker varieties that the seedmen may wish to mix with them. There is much still to be accomplished in securing vigorous varieties of well-known grass seed that will endure drought and shade on both sandy and clay soils. Old fields of meadow land have been found by such investigators as James B. Olcott, of Manchester, Conn., prolific sources of valuable, hardy, and strong-growing varieties of grasses, and the field for discovery in this direction is wide, extending, as it does, from Australia to England.

The operation of sowing grass seed needs to be done skillfully and carefully, and at a time of the day when the wind is light. It is well to sow such seed liberally to make a good lawn. Six bushels of blue-grass seed to the acre may seem excessive, but it will hardly be considered so when we realize how large a portion of any sowing of grass seed will probably fail to germinate. Important as highly tilled ground may be to the preparation of the soil for grass seed, the quicker it is compacted, and kept firm, the better will be the results. It is for this reason the heavy two-horse roller works so well when it is passed frequently over ground that has been recently sown with grass seed, and raked over with a fine-toothed rake. Rain helps still further by also compressing the soil as well as by moistening it. As soon as the grass grows an inch or two, or as soon as the mowing machine will cut it, this operation, frequently applied, tends to thicken the *sole*, or body of the grass, and also, at the same time, to destroy young and threatening weeds.

The frequent cutting of grass in dry weather often weakens it seriously, and allows the vigorous summer weeds to get the upper hand, and therefore, during the first summer, it is a good idea to weed the lawn continually until the grass takes possession of the ground. Care and diligence in watering and weeding a lawn, at this stage of its development, will tend to ensure final results of the most important character. So far as the culture and care of the lawn and garden, respectively, go, there should be little difference made during the first year. Indeed, the author is not sure that it will not pay to give the lawn the most attention, for one may be sure that the lawn will have a long future before it, the success of which largely depends on intelligent treatment in the beginning.

FLOWER GARDENS

THE well-known Hanging Gardens of Babylon, the Garden Parks of the Persians, the villa Gardens of the Greeks and Romans, prove that back in the ages, as far as historical records can help us to reach, the desire in the hearts of men for gardens has been strongly felt and superbly gratified. Since the early periods of history, men have sought to build for themselves homes and palaces with architectural and sculptural adornment; while building temples to their gods, they have not failed to devise gardens for their own delight. Yet it is, nevertheless, a curious fact that the development of the gardening art did not keep pace with the other arts. As far as it is possible for us to know, and we can learn much, since there are abundant records, the arrangement of early gardens was artificially formed, being largely devoid of suggestions of sympathy for nature in any of her free manifestations. And while formal gardens were not as perfect of design in the days of the Persians and Romans as in the celebrated later period of the Renaissance, the beauty of these later villas depended also too much on the architectural adornment; the variety of the trees and flowers was small, and the general arrangement and scheme of the walks and grouping of plants and trees were always artificial. To us men of to-day, who have given serious attention to the subject, it is given to know that gardening cannot be successful where nature is left out of the question.

It is another curious fact that most of our best varieties of ornamental trees, shrubs, and herbaceous plants have appeared, or else have become known, within the space of a couple of centuries. Although roses and other well-known flowers have been grown for ages, the minds of landscape designers were not, except recently, turned in the direction of their best use. While the minds of professional men were working for ages toward the highest as well as the most rational development of sculpture and architecture, the artistic profession of landscape gardening did not exist. No names of landscape gardeners and horticulturists have come down to us from long ago, whereas sculptors and architects have been deified in history and story for thousands of years. To-day, on the other hand, the names of some of the men who grow new plants and design gardens are as well known as those of architects.

Lord Bacon probably realized the relative standing of these arts with great precision when he said, in his essay on gardening, that "A man shall ever see, that when ages grow to civility and elegancy, men come to build stately sooner than to garden finely, as if gardening were the greatest perfection." Whether gardening be a greater perfection or not, the art of gardening in this century has been brought to an infinitely higher state of perfection than it has ever been before, and in the last fifty years we have made such great strides that the forecast of the future indicates the probability of an advancement of the art to an extent of which we can hardly form an adequate conception.

As the number of those practising the art of landscape gardening in an earnest, serious, and dignified way is the best indication of substantial advance, a rapidly increasing gathering of thoroughly equipped men who have devoted themselves with a single mind to landscape gardening can be pointed to in many notable cities of this and other countries. It is because it is a democratic art, one which makes its appeal to the great masses instead of to the few only, that gardening, as we understand it, is in such thorough sympathy with the spirit of our democratic age.

The sum and substance of these reflections bring us to the conclusion that the field of genuine gardening art is so widely open to us that we need not be awed by the great examples of garden-making done in past time, for the more we study, the more we realize that the recorded facts do not bear out the belief that gardening designs have been done in the past as well as they are done to-day, and that the study, therefore, of old examples of the Italian villas and gardens, and even illustrations of much later French and German parks and country places, is becoming less important in the light of the larger theory of landscape gardening which seeks to produce a beautiful, because primarily sensible, arrangement of the present development of flowers, trees, shrubs, and lawns.

In old times they did not have these advantages of superior lawn planting material, nor did they have, apparently, any idea of the importance of making a sensible design for arranging the different beauties and conveniences of home grounds, so that each part would be properly related to the others, and the garden take its duly apportioned place and not absorb the features of the lawn. It should not be understood, however, that precedent is to be undervalued here any more than elsewhere, but it means that we should welcome suggestions from all sources, ancient and modern, plucking from the old masters that which is rational and good. We can and must create, and never servilely copy. We are able to do better work than has ever been done before, not because we refuse to be enslaved by the rules and experience of other days and centuries, but primarily because we have taken to heart the lessons Nature is giving us all about us.

The treatment of each subject, whether a suburban lot, a park, or a large estate, according to the special condition of climate, country topography, etc., *that exists*—this is our fundamental principle, and cheerfully we give up all preconceived and cherished thoughts or theories as to the exclusive beauty of design of the English, French, or German traditional style. The conditions belonging to the house and the surroundings furnish just the right data for inspiration and guidance. Why go far afield for learned examples to copy? We accept all right suggestions, but want no trammels of tradition.

It is this simple instinct for the convenient and fitting that makes the old colonial gardens of New England homes such fine specimens, so nearly perfect in their natural, home-like beauty. It is true, they seem entirely unconscious of any excellence of artistic superiority. They exist where they are, because they are convenient to the house, and do not interfere with the other necessary features of the grounds. The walks are bordered with box, because it is simple and pretty, and easily grown, and the roses and hollyhocks stand where it is convenient to pick their flowers, and where they do not prevent readily reaching the fruit trees and currant bushes which seem to be necessary to old-fashioned gardens; and yet we would not wish our readers to imitate these gardens as they stood a generation ago, and as a few stand now in their sweet wholesome plenitude of charm, any more than we should copy in America the designs of Italian villa gardens, which are altogether attractive where they are, in their environment of classical nature and ancient ruins. We should make our gardens in the way of our day and country. Our great painters do not imitate Raphael or Rembrandt. They speak for themselves out of the fullness of their experience and feelings, and they speak to us, their contemporaries, so that we understand and appreciate them. They are of our day, and talk our language.

Approaching the subject of garden-making in this spirit, we therefore see that the garden cannot be always made in accordance with the dictates of formal theories; neither must it drop into the other extreme and become some kind of haphazard affair of vegetables invaded by riotous clumps of flowers. The vegetable garden, though ever so small, will have its own separate domain, and the flower garden its alloted place. The size and proportion of the space given the flower garden will be governed not only by the pleasure of the house, but also by the amount of space that, in view of the apparent physical and æsthetic necessities, should evidently attach to the lawn and vegetable garden. For instance, a place of half an acre may readily have one-fifth of its space given up to a flower garden in the rear of the house, while on a place of twenty acres or more, one acre would make a large garden, so large that it would need to be isolated and screened in some retired part of the place. There is no reason why we should not get glimpses of a flower garden from points of the landscape more or less distant, but it should seldom be allowed to make a distinct and marked feature of the landscape, with which it is not apt to be in entire harmony. Hence, we should seek to screen it somewhat by a stone wall, masses of shrubs and trees, or some building. Its beauty is much more satisfying if it is seen only when we come directly upon it, and the charming details for which it is specially valuable are at once evident.

An ideal situation for a flower garden, especially one that is made up largely of old-fashioned flowers, such as phlox, hollyhocks, larkspurs, roses, sunflowers, and Black-eyed Susans, is along the edge of a shrub border or piece of woods, where the irregularity of the growth of the flowers and their frequent unsightliness when out of bloom, or fading early in autumn, renders always satisfactory the immediate proximity of masses of green foliage with which they can blend and in a degree lose themselves. A herbaceous border of this kind lends itself more kindly to this type of plants than any other form, and, indeed, for a strictly herbaceous border, the writer doubts whether any other form will be found entirely satisfactory.

Naturally, a flower garden need not limit itself to herbaceous or perennial plants alone, nor to roses, or any other plants that are apt to look unkempt and straggling during the latter part of the summer or early autumn. There are many annuals, such as nasturtiums, asters, pinks, forget-me-nots, pansies, begonias, petunias, zinnias, mignonette, heliotropes, poppies, phlox

EXAMPLE OF FORMAL GARDEN, WITH GRASS WALKS

Drummondi, and marigolds, all of which can be used in this way, if in no other, by planting a succession of late-blooming kinds after the early forget-me-nots and pansies are out of flower. In this way a large garden can be kept supplied with flowers, and the general effect of the open level spaces retained low and neat.

There is, as a rule, a stone wall, or vine-covered fence, or hedges of shrubbery surrounding the garden, and along this can run herbaceous borders of old-fashioned flowers, while in all the central beds will come low annuals and so-called bedding plants (see pages 59 and 65). The effect of this arrangement not only tends to

CORNER OF A FORMAL FLOWER GARDEN

neatness and well-ordered conditions, but it enables the eye to travel unobstructed over the lower-growing beds to the large herbaceous plants on the borders of the garden. The only difficulty with this treatment is the necessity it imposes of erecting some kind of propagating greenhouses where annuals and tender plants can be grown in order to keep up a supply of the right kinds of material for the garden. This does not involve growing plants for greenhouse exhibition, but only a modest supply for the garden.

It is wonderful what a source of interest a modest propagating house of this kind on a half-acre lot may afford the owner by the expenditure of a comparatively small sum of money in the erection of the structure. The writer has prominently in mind an instance where a business man, living on a small lot in a country town, actually floods his rooms with flowers and blooming plants in pots and boxes at all times of the year, growing even rare orchids in this way, besides affording a great number of outdoor garden effects at different seasons in his yard. In this case, the propagating or greenhouse is a primitive affair, the heating apparatus of which seems to almost run itself, as the owner simply shuts off his drafts in the morning after fixing the fire, and in the afternoon comes home to find things in good trim for him to take up his amateur gardening. Amateur gardening it must certainly be termed, for he does not make money by it, but there are, nevertheless, few professional gardeners anywhere who have accomplished such horticultural feats in growing and flowering plants. There has been, without doubt, remarkable skill displayed, the result of long experience and close observation of the habits of plants and a certain genius for making them do what he likes; but, otherwise, in the first outlay of money and succeeding yearly expense and labor, there has been nothing unusual in the surrounding circumstances, nothing that any one, almost, might not have at his command.

In regard to the shapes that gardens should assume, the writer will attempt to give only general principles for the guidance of those who undertake to design gardens, as they should do, in strict conformity with the needs, size, and topography of the village lot or country place they may own. Grass walks or alleys are always agreeable looking and satisfactory, except where dampness for some special reason is feared, or where continual visitation of many people is apt to destroy the vitality of the grass. Walks of gravel that are nearly always dry have a distinct value in the feeling they ensure that one can always use the garden without suffering from dampness.

The edges of gravel paths in gardens should be by preference grass sod, although dwarf box makes a pretty quaint effect. The only difficulty is, if box happens to suffer from winter, or other causes of destruction, it is difficult to mend it, while with the grass border there is no trouble whatever. It is a mistake to make the walks of a garden too wide; it is better to err in the other direction. Four feet is wide enough for most gardens, and six feet is liberal, and yet some broad green alleys of grass may run through the middle of a garden of an acre or two, eighteen feet wide, and look all right with the other walks eight or nine feet.

Gravel walks should be narrower, because they are less attractive than grass walks, and less liable to suffer from the increased wear and tear of narrower walks. The beds, also, depend in shape and size on the dimensions and surroundings of the garden, but they must be narrow enough for one to readily reach and pluck the flowers from one side or the other, say six or seven feet. The beds themselves may be of many forms, but they should assimilate themselves to shapes where there are no sharp points or narrow corners, as there would be in acute-angled triangles. Parallelograms are all right and satisfactory, although they present a modest appearance on paper (see page 64). Forms of the ellipse are excel-

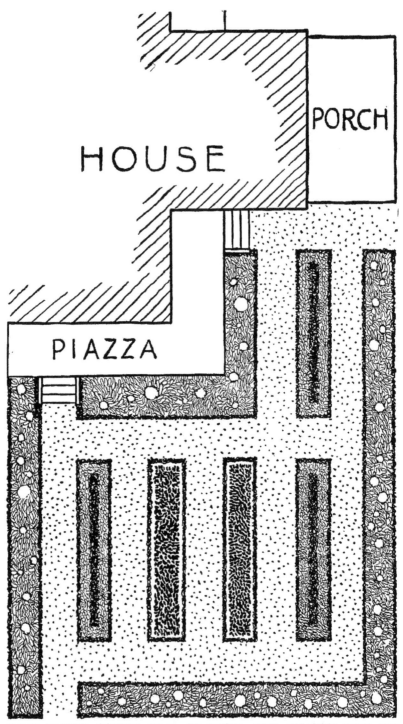

SMALL FLOWER GARDEN, REAR OF HOUSE, WITH STRAIGHT BEDS AND GRASS WALKS

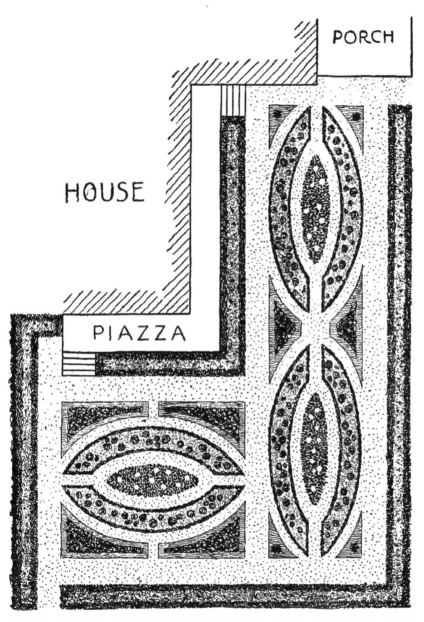

SMALL FLOWER GARDEN, REAR OF HOUSE, WITH ELLIPTICAL BEDS ARRANGED WITH GRAVEL MAIN WALKS AND SUBSIDIARY GRASS WALKS

lent and effective, while the circle, though practically available, seldom fits in as well with an easy and agreeable line of path.

When it is desirable to make a great bed of some kind of low plant, a device is sometimes employed where a subsidiary and narrow path, made usually of grass, is carried around the interior of the large bed, thus securing access for various purposes to the inside of the mass of foliage or flowers (see page 65). Grass is better suited to these narrower minor paths than gravel, even though the main walks of the garden are gravel, and though only low-growing plants are used, as indeed must always be the case with beds thus divided. As a rule, it seems simpler and really more effective to make a bed not wider than six or eight feet, if circumstances will permit.

In selecting plants and arranging them in beds, it is always well to seek to establish them in large colonies, one kind occupying, if possible, an entire bed, or two beds on either side of the path.

Much of the beauty of many flowers is lost when one fails to see them in large masses. It is, moreover, not well to make the effect too monotonous by keeping up everywhere except on the borders the low, even, flat effect. Clumps of Eulalias, symmetrically arranged (page 59), tend to vary the sky line and to please the eye as it wanders over the surface of the garden, and even in each bed a variation of sky line can be produced by using in the same mass several kinds of the one species which differ in height.

There is a system of garden arrangement which is simple and quaint and homelike, and that is a division into large squares, or parallelograms, which consist of plots of grass with six-foot borders running all around them, filled with old-fashioned flowers—phlox, etc.—and at the ends and intersections of the path a single medium-sized choice tree, remarkable for its flowers, such as the magnolia, Chinese flowering apple, horse-chestnut, etc. Seats and summer-houses are used in each corner of

FLOWER GARDEN WITH BORDERING BEDS, INTERIOR GRASS PLOT, AND SHADE TREES AT INTERSECTIONS OF GRAVEL WALKS

the garden, and in the open space of the grass plot one or two fine medium-sized trees may stand, such as the virgilia lutea, the American or European beech, and the hornbeam, birch, or small form of maple, Japanese or Tartarian (see page 68). In such a place may appear an aquatic garden and perhaps a little rustic bridge, all a little outside of the hedge and enclosure of the garden proper. A terrace at one end will also help out the effectiveness of this garden, which should really have all about it outside of the hedge a great grove of shade trees. The same system may be modified for the smaller village lot by keeping a grass plot in the center and running a flower border all around it alongside of the path.

It should always be kept in mind that it is a good thing to distinguish a garden as a flower garden, and not as a garden of clipped evergreens and statuary. Italian gardens are a combination of terraces, lawns, fountains, statues, alleys, and flights of steps, with trees, shrubs, and flowers everywhere. They may be country places, but they are not gardens in the good old homely English sense, the sense in which we want them, if we will consult our own natural and inherited common sense.

THE TERRACE

THE function of the terrace in the convenience and pleasure of the house is not far to seek. It is, so to speak, a sort of exterior living place, an open-air piazza out of which the house rises, and from which it gains a special dignity and convenient opportunity to enjoy most agreeably the open air and changing views made up of lawn tennis, croquet, and social gatherings, which make delightful pictures when framed in holly or box-bordered terrace, allowing one to pass, it may be, from one level to another, and to look from special points on evening sky and landscape.

There need be no confusion as to the intention that should inspire the erection of a terrace. It is not primarily a garden, Italian or French, although it may be appropriately decorated with gay parterres, presenting a generally flat effect, relieved, perhaps, by clumps of grasses, or more formal evergreen shrubs, of which we must allow there are few that succeed in this climate. The ordinary old-fashioned flower, such as the Blackeyed Susan, hollyhock, larkspur, and sunflower, and even the rose, can hardly be used appropriately on most terraces, unless it be on some special part that is backed up by heavy masses of shrubs or woodland. Such flowers are always somewhat straggling and unkempt in their habit during a large part of late summer and autumn, and would seem as much out of place on a terrace as they would set here and there in pots on the piazza. Indeed, the terrace has most of the peculiarities of the veranda, the same simplicity of outline, the same open, severe effect, whereby it affords a more extended and dignified base for the house, and gives it a more permanent and restful appearance. When we see a house reposing on

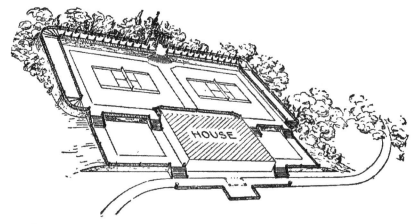

BIRD'S-EYE VIEW OF A TERRACE ON CREST OF HILL, WITH BACKGROUND OF WOODS

broad terraces rising from one level to another, we feel the building has come to stay, in defiance of wind and weather and decaying effects of time, for are not its foundations broad and set deep in the earth? And here, again, we feel the necessity of fitting our terraces to the peculiar and special contour of the ground on which they are to be built.

Terraces need not be, and seldom should be, simply staircases. Their function is rather the broadening and varying the effect of the veranda, and giving the scene a noble and dignified appearance; hence, the terrace finds its most happy situation on a steep hillside where, by its means, all sensation of being perched in air is removed, and a restful feeling suggested as we rise by flights of steps, or by passing out from one story of the house after another, to higher levels. In

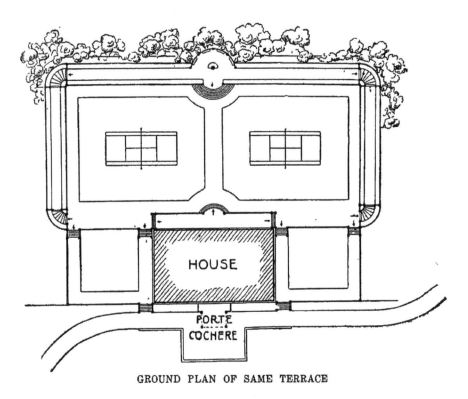

GROUND PLAN OF SAME TERRACE

the peaceful shades of evening one may frequent the terraces on steep hillsides without fear of falling off the edge of things, or wandering rapidly down to spots from which it is difficult to retrace our steps without loss of breath (see page 71).

Level lines of walk with short flights of steps connecting different levels always invite to peaceful and meditative promenade or social intercourse that need not fear interruption by false steps or labored effort. The different parts or stories of the house become more interesting and comfortable looking, viewed from different terraces successively. The house seems to fit itself more intimately to the curves and folds of the land, and, at least in the case of the house of moderate dimensions, to nestle in the hollows of its terraces as if it proposed, in sober fact, to be a hiding place and refuge, a very fortress of protection against all stress of storm and vicissitudes of life. Carrying out the same idea of fitting terraces to the hills, we may curve and fold them into the shape of the ground, never failing to seek the slope of the ogee form rather than the sharp, rectangular, railroad-embankment line, although the rectangular slope has its place, which is generally on more level ground.

But in speaking of the appropriateness of broad, level terraces for steep hillsides, in accordance with the same idea that prompted the builders of the Renaissance villas to hang them on terraces on the hillsides of Italy, we should not attempt to limit the use of the terrace, provided the size and shape of the house is fitted to it. Gentle slopes may lend themselves delightfully to the terrace form, and even on level ground the house may rise with dignity and increased effectiveness from the midst of a broad platform of terraces right and left. Only it may be that the harmony of the proportion of the comparative width and length of house and terrace may be badly adjusted, or it may be that the fair surface of the velvet lawn of the terrace may be disfigured with crude effects of coleus and geranium, or still more marred with improperly used statues and fountains and clipped yews and privets. This unfortunate result cannot be said to be a characteristic fault of the terrace, but simply a fussing and vulgarizing of its proper dignity and simplicity.

It hardly need be said that parterres and statues and fountains can all be used on terraces without injury to their simple and broad dignity, and often with decided advantage to their general effect.

The use of trees or shrubs in the form of a hedge or arbored walk on the highest borders of a terrace also needs attention. Many would use arbor vitæ and privet, with the idea of securing a comparatively permanent clipped formality of outline and height, but in a year or two, perhaps, along would come a peculiar spring, cold and hot by turns, that would destroy a portion of the hedge, and patching would be out of the question; or, in the case of the privet, it would grow naked of stem, especially at the base of the plant, and vexation of spirit for the owner of the place would ensue, not to be solaced by the thought that in England and Europe generally they do this kind of work better with their hollies and laurels and yews, which do not always thrive here.

But the trouble with such workmen is they do not know the material that lies at their hand. If they did, they would turn to the ilex crenata, a Japanese holly, of such hardiness and elegance and symmetry of form and richness of grace of leafage that the only wonder is that any terrace in America is left unadorned by its presence. Pruning it stands well, provided we allow it to be in need of pruning with its perfect and yet graceful and light symmetry; and again it may be repeated, as supremely important, it is hardy in America, for what other evergreen shrub is actually hardy in America, without it be the American holly, a beautiful but somewhat straggling plant, which is specially slow in growth and difficult to transplant as compared with the Japanese holly (ilex crenata)?

So it becomes evident that the terrace is most valuable as an adjunct, a qualification of the house, and may be used with dignified and noble effect in that way; but it needs handling with equal skill of design, although in a very different way, from that employed for the house. The designer should feel the folds and contours of the surrounding ground more completely than is necessary with the house, and above all, the treatment of the surface of the terrace needs to be done with a full and intimate knowledge of the habits of different plants, in such places in this climate. European precedents do

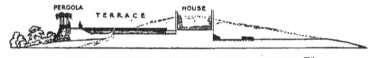

CROSS SECTION OF TERRACE SHOWN ON PAGE 71

not count here, as are indicated by the forlorn attempts we often see to force unfortunate tree box into the clipped shapes of formal gardens. Italy has succeeded with her terrace effects by using plants native to her climate, and by using them in ways that the shape of the ground and the spirit of the landscape suggests. In this same way we, in America, will come to study more intimately our own climate, soil, and landscape effects, and the plants suited to them, before we succeed in securing the best results of terrace adornment.

After all, velvety lawn and open turf effects will always remain the chief charm of American terraces, and the sooner we settle down to accept this undeniable fact and apply it in our landscape work, the sooner we will succeed in developing terraces that will bear comparison with the Italian gardens, that are full of the special character of the vegetation of their country, which it scarcely need be said is not greensward. No picture of terraced hillside and nestling homestead, where we climb from one level to another to find some massive background of woods, could possibly be perfect without rich and broad stretches of turf.

The principle of a terrace in this country and climate,

PERGOLA ON HIGHEST POINT OF TERRACE ON PAGE 71

whatever it may be in others, being strict unity and sympathy with the house and its architecture, it is evident that formality must characterize the design. Flat surfaces and regularly sloping banks, limited to a close association with the house, should in most cases control, instead of indefinitely extended terraces, that may be admissible in a milder climate where life belongs more completely to out-of-doors. It would not be wise to fix any limit to the width of terraces, or the number of them to be employed, and generally, for the sake of simplicity, it is not well to greatly multiply them; and yet it should be remembered that the terrace is not only the narrow strip of raised level ground arranged parallel with the house and laid out along its front, but the whole ground that forms the base or setting for the building. Naturally the terrace may have different forms, from the simple walk, parallel with the house, to the more extensive, massive, and ornate treatment of balustrades and marble-bordered flights of steps, that would properly accompany buildings of more importance.

The treatment of the terrace should have a definite proportion to the size of the house, and to obtain the feeling of security and close relation to the house, a balustrade on the line of some of the terraces may be used effectively; but it seems to the author to be a useless task to attempt to determine beforehand the height and style of the railing, or the width and height of the terraces, so much depends on the architecture and proportions of the house, and, above all, on the special shape of the ground on which the building will rest. We may say in general terms that the greater the depth of slope, the greater should be the distance between the edge of the walk and the slope edge. To unduly shorten this distance is a common mistake; it creates a sense of falling off or insecurity that is inimical to comfort. In the same general way, it may be said that all flat surfaces of the terrace should have a slight fall, say one inch in ten feet, for the sake of proper drainage.

Consideration should be also given the landscape views that might be shut off by some ill adjustment of balustrades and terraces. The effect of the terrace next to the house is to add to its apparent altitude, as the eye insensibly estimates the height from the line of gravel below. Generally, slopes should be made in the ratio of two feet high to one foot wide, and in height they should rarely be over six feet; and where greater depth is required, two slopes should be made, with a level width between them of not less than four feet, and usually a good deal more, for a great multiplicity of terraces is especially to be avoided.

The treatment of flights of steps leading from one terrace to another should be plain and simple, rather subordinating the stone effect by bringing the turf well over the edge of the coping, leaving only such a width of stone exposed adjoining the steps as will be in keeping with the general style of the balustrades and house. In many places of moderate size, no exposure on the upper side of the coping is advisable; simply the effect of a grass border over the inch or two of narrow coping that should protect the steps themselves. The steps should generally be twelve to fifteen inches wide, rising five to six inches, one above the other. There should always be a comparatively level space at the foot of every slope or wall. Care should be also taken to provide for the drainage of the backs of the walls of the terraces as well as of their surface, by means of pipes built through the structure at proper intervals of distance, or by the insertion of a porous backing, from which pipes to carry off water communicate.

PLANTATIONS

IN the minds of many, planting trees and shrubs forms the only part of the equipment of a country place that requires special knowledge and skill, all else being quite within the range of the ordinary uninformed citizen who has not even thought it worth while to inform himself of the peculiarities of his own grounds, much less to seek to teach himself some of the principles of arrangement that should nearly always govern the general treatment of the actual, though perhaps small, surface of the lawn of his place.

If we endeavor to realize at all the true functions of the different parts of our home grounds, or village lots, we will come to make in our minds three important divisions of the work we propose to undertake—namely, the house, the lawn, and the trees and shrubs—and which may impart the most to the beauty and usefulness of the place it will be difficult to say. It is safe to say, however, that the variety of beautiful effects pertaining to the trees and shrubs should occupy no minor place in the scheme of the adornment of the grounds. The general mass may, at first glance, convey a vague impression of charming colors and varying lights and shadows, relieved against the greensward, but as the eye continues to dwell on the details of structure and tint, it will begin to fully realize what an infinity of variety of pleasing effects the small confines of a village lot may be made to develop.

We have seen in the discussion of the best methods of laying out a lawn, that the scheme of its adornment may be reduced to very simple terms, viz.: Framing the boundaries of the place with trees and shrubs, masking with plantations the junctions of the roads and walks, and setting out shade trees along walks and drives; and

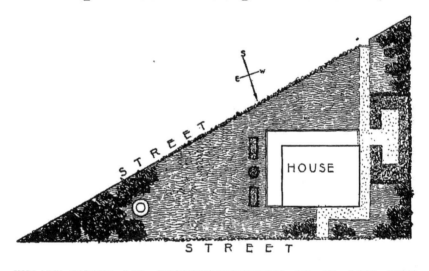

VILLAGE CORNER LOT, FIVE-TWENTY-EIGHTHS OF AN ACRE, WITH BORDERING SHRUBS AND FLOWER GARDEN

this scheme, if the reader will give his mind to it, can be applied to the smallest village yard. Other adornment is chiefly undertaken to set out single specimens of curious and beautiful trees and shrubs as you would erect a statue in the garden, and the result that follows with considerable sureness does not usually, to the mind of the author, make for simplicity, breadth, and beauty of design.

Furthermore, going still more into detail, in the consideration of the exact kind of material to be used in planting we shall find that, in spite of the desire for variety, it is advisable to use a comparatively small number of kinds of trees and shrubs; not because it is necessary to limit variety for the sake of beauty, for then—in a certain sense beauty being infinite and limitless—there could be no limit to variety, but because, when we come to seek shrubs and trees for building the groups on the lawn, it will be found that comparatively few trees or shrubs fulfil the broad and effective requirements of the place.

The shrubs that really perform this office of associating properly on the lawn with their companions of other kinds, presenting both picturesqueness of leaf and flower, and hardiness and vigor of growth, can be counted on the fingers of both hands. The author does not wish to unfairly disparage the charms and useful qualities of many excellent shrubs, but he wishes to point out and emphasize the fact that not many shrubs or trees are actually worthy, on account of general adaptability to all situations on home grounds, to occupy the distinguished position of an all-around good plant for the lawn.

It seems as if it ought to be evident to the reader that if the lawn can be planted with such shrubs and trees as will blend and harmonize and do well together, it would be far better to employ them, even though they may be small in number, than to create a bizarre and unrelated mass of effects with many showy and obtrusive plants that are not in the best sense beautiful in the places they are intended to occupy. It is easy to recognize the truth of this statement when we consider the common lilac, the althea or rose of Sharon, the hydrangea paniculata grandiflora, and a number of somewhat weedy looking spiræas, and come to compare them with the cluster of birches springing from one root, or a mass of hazel bushes mixed with forsythia fortunii or viburnum dentatum.

But recognizing, however, as we do, that few shrubs will assimilate in habit and foliage in the group, we are inclined to employ a number of one kind of shrubs together, depending on a few smaller shrubs, allied in general appearance, to give a dash of desirable variety to the mass; and with this in view, it is a good idea to plant some of these other shrubs singly, or in clusters, just a few feet away on the turf, to give the suggestion of naturalness and informality, so grateful in the arrangement of plants.

It is difficult to explain just why we come to think that certain shrubs look well together. We may try to explain it by saying that the habit and color are similar, but that does not explain it, because it is not true. Perhaps, after all, we shall have to fall back on the fact that we have seen these shrubs stand together on the lawn or in the woods and found that they looked well; but, nevertheless, it does seem as if we would have to refer the ability to make this choice to some instinct that becomes to our minds more seemingly infallible the longer we continue to contemplate different kinds of trees and shrubs.

Having obtained some sensible ideas on the relations of shrubs in groups as to color and form, we must not forget to consider the sky line or top line of bordering plantations of trees and shrubs. A monotonous line of shrubs all in a row and of the same size offends a sense of beauty that very properly seeks variety at all times and places. In lifting up the top line of a bordering mass of shrubs at different points, by planting in their midst at intervals trees of large and distinct character, the value of the general effect of that portion of the group or border thus planted distinctly increases. All definite regularity of these intervals should be avoided, but yet the relations of their several positions will be improved by retaining an intelligently fixed relation between them.

For instance, on a small village lot it may be well to plant elms at each corner of the place, and lindens or planes at intervals that are not regular, but carefully adjusted to some definite theory of arrangement that takes into full account the shrubs that will be expected to grow between them. In the same way, all shrub groups are improved in an artistic sense by setting trees, not at the middle nor in the ends, but at just such places as a carefully thought-out scheme of arrangement will suggest as most natural and beneficial to the general effect.

It is undoubtedly true that the word "natural" may not have a very precise meaning to many, but to such readers the only explanation possible is to refer them to the woods to acquire the proper instinct for this kind of arrangement. The term rhythmic combination is not altogether inapt in explaining the relation that should exist between the trees and shrubs of a group or border.

It should be remembered, in setting out groups of trees and shrubs, that it is not well, if it can be avoided, to plant shrubs, even those best suited for it, in the shade of large, well-established trees, as the roots, as well as the shade, tend to stunt and retard the growth of the last set-out plants. If both shrubs and trees are set out together at the same time, it is another thing, for then neither one will impede the growth of the other for many years.

Before passing to the consideration of individual kinds of lawn plants, we will, in view of the strange ignorance or carelessness frequently displayed by those who use plants, give, without making further excuse for discussing so elementary a subject, a brief space to the subject of tree and shrub planting. The less the subject has been rationally considered, the more do people undertake, without proper thought or study, this fundamental process of the construction of a country place; and sometimes the more gardeners, otherwise skilled, should know about planting as the result of long years of experience, the less ability, whether from actual stupidity or mere carelessness, they display.

The first mistake of this kind, and the most vital one that is generally made, is in the selection of shade trees. The ideal tree for transplanting is, of course, extremely difficult to find in this country, and is scarce almost everywhere, but none the less should we seek to have our trees approximate a high standard in quality and size, for it should be remembered that there is a size that associates itself with the best quality of roots and branches, and which naturally varies almost with each species of tree and shrub. In many cases, people are induced to select small trees when they might easily succeed with larger specimens that would give them quicker effect; and, on the other hand, trees are often transplanted, as already explained, too large to secure the necessary vitality and vigor of root and branch growth for good ultimate results ten years later. The tree that is moved when large, and without the necessary fibrous roots that are found only with transplanting that has occurred within five or six years, will generally live, but will seldom grow satisfactorily, and ten years afterward will perhaps show little

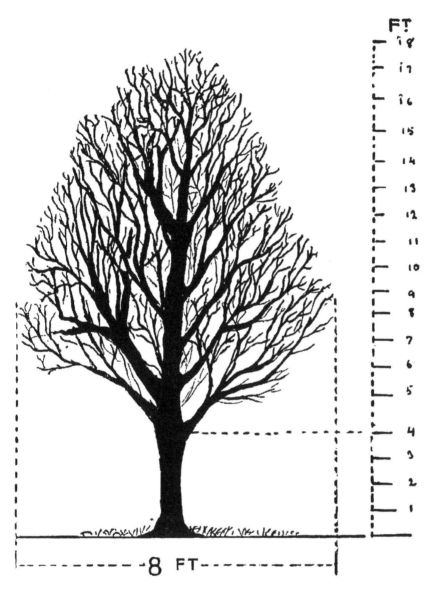

TYPE OF TREE DIMENSION FOR ELMS, MAPLES, LINDENS, ORIENTAL PLANES, AND ASHES, SUITABLE FOR QUICKEST EFFECT AND HEALTHY GROWTH

or no increase in size. For a few of the large and generally popular shade trees, such as maples, lindens, plane trees, elms, tulips, the cut on page 85 will indicate a desirable, though entirely conventional, example of the proper dimensions for superior specimens.

The second mistake we find generally made is setting the plant too deep. It will readily appear sensible to any one who will take time for thought that a tree or shrub when transplanted will naturally like to stand in its new home exactly at the same depth it occupied in the place from whence it came. Some allowance should naturally be made for settlement, which, however, need not cause the mold to be left around the tree more than an inch or two higher than it originally stood.

Recently it was the good fortune, or bad fortune, of the author, for the experience was instructive though the injury was expensive, to see a whole avenue of at least fifty Norway maples, that had been planted by a careless foreman who knew better, set eighteen inches to two feet lower in the ground than they grew in the place from which they were brought; the consequence being death to some of the trees, and to all a check in the growth that it will take years to overcome, although they have been, for some time, raised to the proper elevation with relation to the surrounding soil.

Furthermore, the hole should be dug considerably larger than the area of the roots of the tree when spread out, and should, moreover, be made considerably larger, and not smaller, as is usually the case, at the bottom than the top.

In regard to the necessity for improving the character of the soil that is to be used in the hole, especially in light, sandy ground, the author feels that it is impossible to express himself too strongly or with too much gravity. It may seem to the ordinary observer that excessive expense is demanded when twenty loads of rich mold are recommended for filling a hole for a shade tree in many parts of cities like New York. There are, however, many regions where it would be profitable to dig the holes for trees ten feet in diameter and four feet deep, and there are very few places where it does not pay to dig a large hole for a tree and fill it full of fresh mold, and the same treatment is, of course, advisable for shrubs, herbaceous plants, and vines.

There is a sort of painstaking care that a good planter exercises (and there are not as many good planters, by any means, as there are professional gardeners), which makes him fall on his knees and work the fresh mold among and under all the fibres of the larger roots, and then shake the trees sideways, and up and down, until the spaces between and under the roots are thoroughly filled with fine earth, the final process being the tramping or ramming the earth around the tree, layer after layer. All this compression of the earth over and around the roots tends directly to keep the air away and start the growth with desirable quickness and vigor.

In many kinds of soil it is profitable to prepare a special system of drainage, and a convenient method of watering each tree. For a succinct and clear review of the subject of this thorough and specially successful kind of tree planting, the author would refer the reader to several recent reports made on this subject by what in Paris would correspond with a bureau of planting, if we had one, in the Department of Public Works in this country.

After the trees and shrubs have been planted, the

practice of cultivating and watering them, which is unfortunately often neglected, should be at once encouraged, to prevent death from drought or hot suns, or perhaps only the inception of a stunted condition that many years will often fail to overcome. Tree guards, and stakes for keeping the trees firm and straight in high winds, should be duly set as soon as the operation of planting has been completed. It is a good idea to raise the beds or territories in which trees and·shrubs are intended to be grown, six inches to a foot, for the purpose of good drainage and the better display of their attractions in mass.

It has been already intimated that herbaceous plants that grow and die down every year may play a special and most important part in the general harmony of lawn effects. Their proper place is generally found in irregular borders along the front of shrub groups and in the angles, or along the base of the house or other outbuildings, and at the foot of walls and fences.

In presenting a statement of the proper relation of the different plantations on the lawn, we would say, first, grass; next above, herbaceous plants, and then shrubs of a comparatively small size; and, finally, large shrubs, with trees at considerable intervals throughout their mass. An arrangement of hardy plants made in this way will be seen, on experiment, to be more harmonious and effective than either of the classes employed would look if set out alone.

In many places, both in public parks and private grounds of small size, and especially in those located in rural-looking territories and spots, the substitution of vines and low shrubs in place of ordinary grass borders may be practised successfully with economy and excellent effect. If the vines—climbing roses, honeysuckles, and Virginia creepers—are cut back or thinned out a little once or twice a year, to prevent disorder and encroachment on the footpath or roadway, the effect will prove both natural looking and agreeable.

On highly cultivated and more evidently artificial grounds, the effect of vines and small shrubs bordering the roadside seems to the author to look somewhat forced and out of place. It scarcely needs to be repeated that herbaceous plants find their best and most congenial home in the regular old-fashioned garden set aside for the purpose.

Bedding or foliage plants are not hardy, and are consequently set out freshly every year in formal beds specially designed for the purpose. They consist of such plants as the coleus, geranium, and canna, and are alien, exotic productions when they appear on the lawn, where they consequently do not seem to be at home except immediately adjoining the house, where they become, very properly, part of the domain of the architecture. For the same reason, they become an appropriate adornment of the city square or public or private courtyard.

It is not at all the wish of the author to in any way discredit the beauty and value of bedding plants appropriately placed, but simply to ask that they shall be duly coördinated with other plants and the surroundings of the picture, and, above all, to seek for them the application of the same broad artistic principles of design that we have advocated for the arrangement of trees and shrubs. In a miniature way, there should be just the same principles applied whereby, in place of tall trees, will appear cannas, palms, musas; in place of large shrubs, acalyphas and geraniums; and finally, in place of the low bordering herbaceous plants, alternantheras and pyrethrums, and the whole will then become an aggregation of harmoniously and picturesquely designed parts.

With such planting material and intelligent designing ability, endless beautiful combinations may be worked out, on lines as bold and free in their way as those used in the large shrub group, and sky lines as sweeping and noble, and arrangements of colors as soft, delicate, and harmonious. All strictly flat treatment of such foliage beds, in imitation of rugs and other purely formal and artificial designs, have an element of the meretricious and vulgar about them, which, however brilliant and intricate the pattern, will be likely to strike a note in the general scheme of harmony that will not be pleasing to many who, perhaps, may not exactly know the reason why.

It is certainly not necessary to say that flowers add greatly to the beauty and pleasing effect of foliage beds, but if used in too large quantities, they will mar in the first part of the season by the insignificant appearance of their leaves, and later on by the overshadowing and varied mass of their individual bloom, just as the illustration may be seen any year in some foliage bed where a predominant grouping of white lilies is often so widely spread throughout the mass as to largely destroy the intended effect of the design.

The desire of many in regard to the distance at which trees and shrubs should be set apart is turned a great deal in the direction of securing immediate effect. It may be that one is justified in undertaking to plant for immediate effect, provided he has entirely counted the cost and expects to thin out, after a few years, many trees and shrubs annually. In actual practice, however, in case of close planting, experience has proved that, as scarcely any one systematically thins out plantations so as to allow the remaining trees and shrubs to properly develop their normal character, it is a good idea under the circumstances to take a conservative course, and plant at distances that will allow the trees and shrubs to develop properly for a number of years without injuring each other, and at the same time to keep them near enough together to look well even when they have attained a moderate size.

It is difficult to fix the exact distance apart that trees and shrubs should stand, because the growth of trees and shrubs, even in the case of the same species, varies to a remarkable degree, as all close observers will remember; but we should say, in a general way, that large trees, like elms and maples, should stand forty to fifty feet apart, and shrubs like dogwood and snowballs, six to eight feet apart. This will allow a fine exhibition of the characteristic beauty of the plant, before the time comes when the thinning-out process must commence, if the value of the trees and shrubs is to be saved from nearly uniform deformity and decay. The effect of close planting is almost invariably in actual, as distinguished from theoretical, practise, to crowd more or less of the plants before the pressure of the overgrowth is relieved by cutting out or transplanting.

It has the tendency to simplify the problem of planting, and the accompanying thinning-out process, if large masses of one kind of shrubs or trees are used, with perhaps on the outskirts a slight sprinkling here and there of other kinds, to prevent a sense of monotony; for in that case it becomes a simple thing to select a number of plants for removal from a quantity of the same kind. It will be found that this is a more rational process than the one that undertakes to use quantities of shrubs as *fillers* between specimens of trees and shrubs set wide distances apart, and intended to remain permanently after the filling shrubs have been taken away entirely. The difficulty of this plan, which seems plausible at first sight, and has its good side, is that the growth and fate of neither the permanent plant nor the filling plant can be foreseen, for any one of them may grow very much larger or smaller than could be expected, or may die, when much of this particular theory of arrangement would come to grief. If fillers are used, low-sized shrubs and vines, such as azaleas, itea virginicas, hypericums, honeysuckles, rubus hispidus, and roses should be employed.

So far as the preparation of the soil goes, in spaces intended for masses of trees and shrubs they should be spaded up and cultivated in large beds, whether the groups be planted thickly or not, because in this way the healthy and symmetrical growth of stem and foliage is so much more benefited than if they are set in holes dug in the grass, as is generally the fate of shrubs planted some distance apart from each other. In a few years it will be found that the plants will, in spaded beds, completely occupy the ground, and render it unnecessary to cultivate them.

It is necessary, in arranging trees and shrubs on the lawn, that the situation of each shrub shall be chosen, not only for its own individual exhibition of attractive qualities, but for what should be the dominant idea, of preserving its proper harmonious relations to the general mass of which it is to form a part; for, after all, the chief function of its existence must always be to contribute to the value of the lawn as a genuine picture.

DECIDUOUS TREES

FOR the various beauties and excellent qualities that should pertain to a good tree, the sugar maple should be accorded the distinction of ranking first in the maple family. It grows well, and is long lived, and not difficult to transplant under most conditions, but it must be conceded that it does not succeed in all kinds of soil, not even in some soils in which other trees do well. In the sandy soil of New York and other great cities it does not grow as satisfactorily as, from the general behavior of the tree, we have a right to expect. Up the State, in New York and in New England, and largely throughout the entire North and Middle States, its beauty and thriftiness are remarkable.

For general usefulness in lawn planting, the Norway maple, although it does not equal the sugar maple in actual beauty and symmetry, is one of the best trees we have. It will thrive in all kinds of soil, and do well in all exposures; is free from disease, attains noble size, presents a great rounded mass of foliage, and affords excellent shade. To see the Norway maple at its best, is to contemplate a specimen that branches so low as to have its twigs and leaves resting on the ground. One of the finest trees of this kind the author has ever seen was in Baltimore, Maryland, and in this case the massive foliage swept the ground, and from thence rounded into a great sphere of green seventy-five feet high.

Both the sugar and Norway maples retain the natural beauty of form for which they are celebrated, without pruning, except for the removal of a dead limb or abnormal shoot. The Norway maple blends well with other trees, and has less of the prim symmetry that renders the sugar maple less suited to a congregation of different species. It is, therefore, an excellent idea to group sugar maples in masses by themselves, or to scatter them singly throughout a particular territory.

The sycamore maple is a less symmetrical tree than those just mentioned, but its foliage is fine, and the appearance of its seed vessels attractive. It is hardy, healthy, and easily transplanted, and naturally develops reddish purple varieties.

The scarlet or swamp maple has considerable value, but it is not of equal excellence to the sugar and Norway, being loose-headed and irregular in outline. Its proper habit being the swamp, it is naturally not suited to dry, sandy soil, but when it does thrive, as it readily does on ordinary soil, it is a fine tree. In autumn its coloring is splendid, and in spring it has charming red flowers.

Although it may be said that the family of maples all afford satisfaction on the lawn, there is possibly one exception, and that is the silver or soft maple. It is true it grows fast, transplants readily, and is attractive, sometimes, for a number of years, but eventually, and sooner rather than later, its brittle branches are broken down by ice storms, the foliage becomes sparse and comparatively uninteresting, and, too often for the credit of the tree, early decay sets in. Its form, at its best, is seldom specially symmetrical or graceful. Having no really remarkable characteristics, it is less attractive and valuable than some of the quick-growing, short-lived poplars which have, without question, striking forms and fine foliage.

Many soft-wooded, rapid-growing trees become popular just because, by their employment, the desirable landscape effect can be quickly obtained. This quickly-obtained effect has doubtless advantages that we cannot afford to disregard, but there is a great difference between rapid-growing trees, and it behooves us to see that our selection does not include brittle, early-decaying soft-wooded trees, like the weeping or Babylonian willow and the silver maple. The effective Oriental plane tree and the American elm, on the other hand, are also rapid-growing trees, but at the same time long lived and clean and solid in structure.

Chief among the most interesting and choice maples are the various forms and colors of the endless varieties of the Japanese maple polymorphum. Generally these maples are small, not larger than a shrub, and in some cases they are actually dwarf. They are popular with the Japanese, and appear frequently in the designs of their artists. Other excellent maples from Japan are of large size, but they are so rare in the nurseries that they can hardly be said to form a part of our lawn-planting material. There is a Tartarian maple, and also an English field maple (acer campestre), both of which are valuable on the lawn, owing to their comparatively small size and picturesque and, in the case of the former, fine autumn foliage.

The American elm, as we have just pointed out, affords an illustration of a soft-wooded rapid-growing tree that often lives to great age, and to the last presents great attractiveness of outline. There is nothing quite so fine in general contour, among the denizens of the lawn, as the dome-like shape of the American elm. As we look down the streets of some of the New England towns, the vistas through the arching American elms remind one of the majestic aisles of some old-time Gothic cathedral. Besides growing fast, and presenting in mature age a noble appearance, the elm is capable of adapting itself to different kinds of soil, both wet and dry, which is surpassed by few other trees. Yet it is a tree that readily responds to good treatment, and likes plenty of rich mold and moisture at its roots.

The English elm grows to be a tree of noble properties and rounded, massive shape, but it is entirely unlike the American elm, making an altogether different effect in the landscape. There are several other good elms in common use on lawns, notably the cork-barked and Siberian kinds.

In the same rank as the maples and elms stand the lindens. Of large size, generally symmetrical form, rich green foliage, giving abundant shade, and, above all, displaying a rapid-growing, easy-transplanting faculty, the several varieties of lindens are of unsurpassed value on the lawn. The English or European linden is especially symmetrical and noble in appearance, but the American species, while somewhat less interesting in coloring and outline, is of quicker growth, and adapts itself more readily to all kinds of soil and exposure. Like the Norway maple, the linden should be encouraged to grow its limbs close to the ground, and the silver-leaved variety of linden displays attractively the white undersides of its leaves to the breeze. Unfortunately, disease and insects do, after all, attack the European linden during some seasons, while the American species is comparatively free from pests of all sorts, thriving on the poorest soil in the most crowded parts of the city.

Of all the sonorous-sounding names of horticulture, the most impressive is the liriodendron tulipifera, the familiar tulip tree of American woods. Yet familiar as it is, and in spite of its somewhat high-sounding name, it is a precious possession for all tree-lovers, who fully realize that for beauty, dignity, and general effectiveness it has few superiors among shade trees of the largest-sized class. It is almost always found to be healthy, vigorous, and rapid-growing, after it has been for a year or two in the ground, for, like the magnolia, it has roots of a spongy texture which are easily injured in transplanting, and made to suffer on account of sensitiveness to drought and the sudden changes of winter and early spring. For this reason, it has been found better to plant tulip trees in the spring rather than the autumn. The great rounded trunks of mature tulip trees, rising in some specimens fifty feet in the air before the foliage begins, are particularly impressive. The foliage seems to recede as the years pass, for in youth the mass of leaves is large, and spreading and dominant. The individual leaves are of a fresh and lovely green, and most curious and elegant in shape, and in autumn they often assume an attractive golden hue.

Pruning seldom needs application to the tulip tree, except for removing a dead limb; indeed, the author is inclined to wish that the use of the pruning knife could be totally abolished, for the injuries it perpetrates are far greater than the good it does. There are few trees, or shrubs either, that do not lose their most characteristic beauty of habit by the ordinary system of pruning. The essential and characteristic quality of a tree or shrub, by which we are accustomed to recognize its special and abiding charm, is apt, in many cases, to lie in the tips of the branches, and against these tips the pruning knife generally wages war, dire and effective. Perhaps we would not go far wrong if, instead of abolishing the pruning knife, we limited its use to only dead or diseased wood.

Being a relative of the tulip tree, we need not be surprised to learn that the magnolia has still worse transplanting qualities, and all that has been said about capricious habits and slow growth during the first years of its life on the lawn, especially if planted in the fall, applies with much greater force to the magnolia, and goes far to explain the reason why we do not see the latter planted more generally. Yet nearly all the species and varieties are beautiful and striking, both in flower and foliage, and worth the most painstaking efforts to establish. Among the specially beautiful flowering kinds we may mention the Japanese sweet-scented white magnolia stellata, earliest blooming of its family, and also the early magnolia conspicua, magnolia soulangeana, and magnolia lennei; and among the June-blooming kinds we have the richly-scented, cream-colored, red-stamened magnolia hypolenca, and magnolia watsonii.

For magnificence of foliage no tree in the North can surpass magnolia macrophylla, its leaves being eighteen inches to two feet in length, and its great white flower as much as a foot across, while its size often reaches, under favorable circumstances, forty feet. All the magnolias, it should be said, have fine foliage.

As we turn over the pages of classic history, we find

that the Oriental plane, above all trees, seems even in those early days to have surpassed other species in longevity, tales being told of specimens living more than a thousand years; and yet this tree presents the apparent anomaly of being, in spite of the fact that quickness of growth and shortness of days in arboreal life often go together, one of the most rapid-growing and easily-transplanted trees that we can employ on the lawn in either country or city. In Europe, and all over the East, it has long been highly valued for lawn planting, but in America, of late years, it has been undeservedly neglected, although appreciation of its value is again growing rapidly.

From the very start, the plane tree grows vigorously, and as the years go by its vigor never flags nor its foliage fails to grow large, umbrageous, and healthy. In old age the trunk is disfigured by scaling bark, and, on the American plane especially, the top of the tree sometimes dies, an accident, however, which may, perchance, happen to any rapid-growing tree, like the American elm, for instance.

If it were necessary to seek for a more hardy, vigorous tree, under adverse conditions, than the plane tree—and it would be hard to imagine that it would be worth while—we would, perhaps, turn to the honey locust (gleditschia triacanthos). This tree has graceful foliage, and a dark, attractive-looking stem, covered with thorns, which is a drawback, it must be confessed. Its light shade, moreover, renders it scarcely fit for a street tree. The chief value of the honey locust, though otherwise a fine tree, after all lies in its wonderful adaptability to all soils, which enables it to thrive in the sands of the seashore and the slums of a large city, and to cling to the scanty soil of a cleft in a great rock, and thrive therein most luxuriantly. There is a kind of honey locust (gleditschia inermis) that possesses the great advantage of being thornless.

We have already spoken of the poplars as affording an instance of a quick-growing family that is apt to lose its beauty, and even die, in a comparatively short space of time, but we are not prepared to give up its use, because groups of the different species may be so disposed as to produce an immediate effect that will last for a time, and can then be replaced by neighboring and more permanent trees. The period of the beauty and vigor of poplars may be extended by removing dead or diseased wood as soon as the smallest amount of it appears. The Carolina poplars and balsam poplars are good kinds, and free from the objection of suckering, which has so greatly injured the reputation of the silver-leaved species.

For landscape effect, the Lombardy poplar is the most valuable, pointing, as it does, its spire-like form far above the general mass of surrounding foliage. Pruning away dead and diseased wood is especially valuable in the case of this tree, because it tends to renew a fresh, marked vigor of growth. At the corners of small places, on either side of a gate, or along the side or back of the house, the Lombardy poplar produces excellent effect in the landscape, but it should always be associated with large shrubs or other trees, as the lower portions of it are apt to be bare and uninteresting.

Many of our native trees are both beautiful and well suited for the lawn, and ranking high among these we find the yellow-wood of Kentucky, virgilia lutea, or, more properly, cladrastis tinctoria. It is a charming tree, slow in growth and of beautiful, refined nature. Every curve of the tree is fine, the smooth-barked round trunk, the quaintly shaped twigs and cheerful-looking light pea-green leaves. The white flowers are apt to bloom biennially, and are very attractive, reminding one of those of the white wistaria with its long, loose racemes or clusters.

Memories of spring are always rendered more delightful by visions recalled of apples, peaches, and cherries in full bloom; indeed, many would be inclined to allow that no pleasanter sensation is produced by any tree than that made by the rounded form of the apple tree in flower. It is fortunate, therefore, that the horticulturist has been able to grow varieties of these fruit trees which, instead of developing fruit, reserve all their vigor for the purpose of producing greatly increased numbers of larger and more beautifully-tinted blossoms. It has been thought worth while by tree lovers to travel hundreds of miles to enjoy the rare treat of beholding a fine specimen of Parkman's double-flowering Japan apple in full bloom, and we are inclined to believe that the reader, if he could see a good specimen of this tree in flower, would agree that the time occupied in much travel, for the purpose of seeing it, had not been ill spent.

The trouble with the horse-chestnut, although it is celebrated for its flowers and foliage throughout the known world, is that the time comes in July and August when, in America at least, the beautiful delicate green crinkled leaves turn brown, and fall long before the foliage of other trees.

The mountain ash and ash-leaved maple are subject to attacks of insects and disease, and have a certain coarseness and looseness of habit that have tended to render them unpopular, and the catalpa, though rapid-growing and effective, on account of the shape of its great leaves, has an ungainly appearance that counts against it. One large-growing ash there is that apparently continues free from disease and other drawbacks which render many other ashes unsatisfactory on the lawn, and that is fraxinus Americana, the American white ash. Its foliage is rich green, and in form it is symmetrical. It does better here than the European species. The European larch is charming in its early growth, but later on in summer it often grows rusty.

One of the most picturesque and generally valuable shade trees is the odd-looking American species, with glossy, star-shaped leaves, turning red in autumn, and surmounting a rough-barked, attractive trunk, and bearing the euphonious, smooth-sounding name, liquid amber. It is not rapid in growth, nor especially easy to transplant, but it eventually develops into a large tree, and affords agreeable shade along any road on which it may be planted.

Poets have many times sung the praises of the dainty and graceful white birch, "the lady of the woods," so we will confine ourselves strictly to its practical advantages for the lawn. Being of medium size, its somewhat pointed and very marked contour serves well to vary the sky line of any group of trees and shrubs, and in winter, or against a background of evergreens at all seasons, the white stem makes one of the most notable effects in the landscape. The birch is a capricious tree in many ways. It will take a fancy and stop growing, and then start in and grow rapidly, and again it will occasionally die unaccountably, and in transplanting it will also act queerly at times. The best time to plant it is in early, not late, spring, and surely not in fall, although the reader can undoubtedly point to successful fall planting of birches. And yet we must have the birch, for in spite of these drawbacks it is without doubt one of the most precious and important elements of the landscape, whether we see it singly growing, alongside a great rock or against a background of evergreens, or in clusters springing from one general center. In order to secure success, and it is not really a difficult tree to transplant if care is taken, we should see that the young fibrous roots are not injured in digging or allowed to dry up during transportation to the place where they are to be planted. The happiest situations for birches are along the shores of streams, or in the midst of shrubs on the smallest village lot, where they present a most distinguished effect, and serve to vary the outline and sky line of the bordering shrubbery in the most delightful manner.

No one, I fancy, will dispute that the beeches are among the most richly endowed of shade trees. They have great longevity, and exceeding beauty of trunk, branch, twig, and leaf. The trunk has rounded contours and greenish gray tints that are attractive in all species and varieties, whether European or American. The twigs are sharply pointed and quaint, especially on American beeches in the winter time, and the leaves are rich and glossy, and group themselves in effective masses.

Fagus ferruginea, the American beech, has always seemed, for some unaccountable reason, to be neglected. It is extremely interesting with its light-gray bark and layer-like masses of elegant-looking foliage.

Purple beeches and weeping beeches are renowned for beauty, but they should be used sparingly, as their somewhat abnormal appearance, beautiful as it is, partially unfits them for blending harmoniously with the general mass of trees and shrubs on the lawn. As a rule, beeches do not transplant easily, and should be, therefore, set out small, four to six feet high. Beech hedges present a picturesqueness and variety of outline that go far to redeem the general stiffness and monotony of the hedge when compared with the varying play of form and color that characterize a well-selected and grouped border of shrubs and trees. The hornbeams possess many of the good qualities and attributes of the beech, and apparently thrive in the poorest soil.

It is difficult to classify trees, and say which is comparatively best among the modifying circumstances of soil and exposure, but for general excellence we feel forced to rank the oak very high, as high, perhaps, as any. It is conceded on all sides that oaks are sturdy, enduring, and picturesque, both in rough bark and glossy and curiously indented leaf, but it is objected that they are slow of growth and difficult to transplant. Generally it is the fault of the planting and quality of the soil used that retard the growth of an oak. It is surprising to see how closely a pin oak, when it has been carefully planted in rich soil, will keep in size to a linden or maple, and as for the difficulty in transplanting, all we can say is that it usually grows out of the fact that the specimen in question has not been moved for the purpose of getting new and abundant fiber later than five or six years before the last date of planting. It is likely, moreover, that it will be found, in case of failure, that a large stunted specimen, ten to twelve feet high, has been used, instead of one four to six feet in height that has been recently transplanted.

The pin oak, one of the best growing kinds, is considered valuable on account of its finely formed dark trunk and drooping, graceful, deeply indented foliage. The red oak has rich red leaves in autumn, and the chestnut oak and black oak are also excellent. The American chestnut is the most ornamental of the nut trees, and grows to great size, the foliage being large and glossy, and the flowers in June white and fringe-like. Its only drawback is a certain shyness in transplanting.

Strangely constituted, and yet one of the best and most attractive of Japanese trees, the Japanese gingko (salisburia adiantifolia), or maiden-hair tree, has long, spreading branches, reaching out in a peculiar fashion of its own, and bearing cones when it fruits, which seldom occurs, for it is unisexual. The leaves are light green and curious, resembling in outline an open fan, and are borne in somewhat sparse numbers along the branches. The gingko is entirely hardy, and grows eventually to be a large tree, forty feet high, though in youth its growth is not rapid.

The sophora is another hardy Japanese tree, bearing in June cream-colored pea-blossom-like flowers in racemes, and, like the gingko, attaining large size in a decidedly slow fashion.

It is a pity to feel obliged to leave the subject of deciduous trees with the full knowledge that it has been necessary to pass unnoticed many excellent shade and lawn trees, but it has been the endeavor of the author, within his limited space, to discuss at least a number of the most distinct and best trees, for beauty and growing qualities, that are likely to be found in reasonable quantities in the nurseries of the country.

DECIDUOUS SHRUBS

IT is rapidly coming to be an accepted theory that American trees and shrubs should, to a great extent, inhabit American lawns. Some persons go so far as to almost assert that nothing but American trees suit American lawns, but the continued employment of such trees as the Norway maple will serve to disprove the theory as one to be universally applied. However, be that as it may, there is no doubt that many hardy native shrubs grow remarkably well together, and blend their several attractions in a thoroughly harmonious and natural manner.

Ranking high among such American shrubs comes the common alder of the fields and woods, alnus incana. It is an early spring-blooming and swamp-loving plant, and thrives better in wet places than almost any other shrub. The early effect of its satiny leaves pushing out when little vegetation has as yet commenced to move, makes one of the chief beauties of spring, and in late fall, on the edge of winter, a few yellow flowers will already appear on the alder, as if it were the intention of the plant to announce, in a slight and uncertain way, what it proposed to do, in a full and effective manner, as soon as spring appeared. Alders transplant and grow well, branch symmetrically and compactly from the ground upward, forming natural companions for the willows.

Willows appear with their catkins, or blossoms, in early spring, and take almost equal pleasure in wet ground. They grow with great rapidity, and even better than alders, thrive on almost any kind of soil, although as they grow old they become coarse looking, and are apt to fall into early decay. This applies especially to the weeping willow, salix babylonica, the long soft shoots of which soon begin to droop in a forlorn and distressed condition after it has attained the size of a tree and ice storms and other accidents encounter its brittle wood. It is evident that it is only in extreme youth that the weeping willow exhibits any shapely vigor or positive charm. Bushy forms of willows, of which there are legions, last longer than the weeping willow, and perform excellent work in accomplishing attractive effects in the landscape. The yellow-stemmed willow (salix aurea) is, perhaps, the best of these bushy forms, growing in great clustered masses in almost any soil, and, although naturally liable to become coarse in time, is likely to thrive well for a long period of years. Laurel-leaved willows (salix laurifolia, or pentandra) are less interesting, and grow less attractively and more stiffly than the golden kind. Their foliage is very fine, resembling that of the orange tree.

For such large-growing shrubs plenty of room is required. On small plots, owing to the limited amount of space, willows of this character should be planted at the extreme rear.

There is a wide range for effect in color in autumn among American trees and shrubs, and chief among the latter may be ranked andromeda arborea, the sorrel tree.

Its leaves shine with a rich red light in the western sun of autumn, making what some think is perhaps the best effect of its kind to be seen on the lawn. The large, fine white flowers come in July, and until winter the foliage is a shiny green. In association, the andromeda arborea looks well with azaleas, kalmias, and several other kinds of andromedas, although it grows larger than any of them. Noteworthy among the other kinds of andromedas are catesbii and floribunda, all smaller in size and earlier blooming than arborea.

One of the most charming families for any lawn, provided it has a suitable place for its growth, is the hardy azalea of our American woods. Azaleas should be set on a hillside, or in a ravine, where a natural woodland effect can be contrived by backing up and framing the plantation with plenty of large trees and shrubs. With the masses of azaleas (and they should be disposed together in large quantities by themselves) will naturally grow American shrubs like callicarpa, ceanothus, clethra alnifolia, rhodora canadensis, and itea virginica. Foliage and flowers are all fine on these shrubs, and the coloring in autumn on several of them is especially rich.

In spring the earliest positively glowing effect of flowers is found on the closely set, bright-red blossoms which come on the Japanese quince (cydonia japonica). The picturesqueness of the foliage of this shrub adds beauty to its appearance, because it acts as a foil to the numerous flowers, but disease sometimes assails it badly, and it certainly does not thrive as well in all kinds of soil as some other shrubs.

Spring's most characteristic and entirely satisfactory shrub is the white-flowering dogwood (cornus florida). For flower, foliage, and picturesqueness of outline there seems to be nothing that can surpass it at that season of the year. The foliage lies in stratified masses that are very effective, and the leaves are finely formed and beautiful in detail, as well as in the way they bear the simple and broadly outlined flower. The very tint of the bark is fine, and in autumn every one knows how much our woods owe to them. Another fine cornus is mascula (cornelian cherry). Like the C. florida, it comes early with yellow flowers, and in autumn has bright-red berries. Other corni—stoloifera, siberica, alternifolia, and circinata—are fine in foliage throughout the summer, but they are not remarkable for their flowers nor for their autumn coloring. The red-stemmed dogwood is hardly less valuable than the white-flowering species, for its red twigs are particularly effective in winter, and in summer the growth is vigorous and picturesque above most shrubs on the lawn.

Deutzias are well-known and popular shrubs, but they have a formal, stiff habit that suggests the idea that they belong to the garden rather than the lawn; and weigelas and lilacs and altheas all give one a similar feeling. It admits of no question that in full bloom they are beautiful, but during the remainder of the year their appearance does not blend well with other shrubs on the lawn. The Japanese and Chinese lilacs are less open to this criticism than the common kind, vulgaris.

It is strange that the hazel bushes, both American and European, are so little used on the lawn. Their leaves are interesting in shape, and the habit of the plant is compact and picturesque and effective in masses, whether standing alone or in combination with other shrubs. It has a permanent look and great vigor united with much refinement.

Almost as popular and well known as the white-flowering dogwoods, the forsythias always constitute, whereever they grow, one of the most conspicuous effects on the lawn. Their yellow flowers grow in great, close-set masses. Forsythia fortunii has the most effective foliage, because it is larger and more characteristic and satisfactory, and blends better with other shrubs.

An effective Japanese shrub is elæagnus umbellata, and its companion, E. longipes. It is a great, rampant-growing plant of picturesque shape, and has leaves with silvery undersides, and small yellow flowers followed by bright-red fruit. Elæagnus hortensis has beautiful silvery, grayish-green foliage, but its habit is less vigorous than that of the two species first named.

Euonymuses are a little stiff and tree-like to serve the purpose of a shrub, but they have a fine bark and habit of growth, and their numerous, curiously shaped red seed-vessels make a striking feature in the landscape on a sunny autumn day. To these attractions are added, in the case of some species, a rich fall coloring, that is well illustrated in euonymus latifolius, E. atropurpureus, and E. alatus. The green of the euonymus leaf often exhibits a fine bluish-purple hue.

It is not the intention, however, of the author to advocate purple and golden-leaved kinds of any shrub or tree, notwithstanding the fact that they are universally popular. Viewed individually, such plants are undoubtedly beautiful, and have their value for those who may desire to multiply their specimen plants; but for the landscape they present, as it seems to the author, a quality that is not desirable in the group, because it is essentially scenic, as opposed to the natural, in the striking glow and glitter of its display on the lawn.

In the case of shrubs and trees beautifully colored in autumn, this scenic effect merges easily into the natural, because the purple and red and gold are mingled with the green, and all nature, in dropping her royal robes, runs through the entire gamut of color from glowing crimson to dull browns and grays, and consequently no bizarre, inharmonious, and abnormal effects are suggested, any more than they would be in the weird changes of decay and death in other portions of Nature's organism.

The barberries furnish an illustration of the superior value of the green over the purple, and, strange to say, the very habit of the green form is less stiff and more graceful than the purple; the abnormal color seems to predict a certain stiffness, dwarfness, or otherwise unnatural variation of the original form of the plant. Smallness, on the other hand, does not imply dwarfness, which conveys a distinct sense of the abnormal, for the best, and certainly a quite natural looking barberry, is the Japanese species, berberis thunbergii; the entire plant is constructed on a small but very effective scale. Its height at maturity, generally in this country only four or five feet, is moderate; its leaves, flowers, and berries are all small, and the shading of its glossy leaves, from June to October, through the varying tints of green and deep red, is refined and delicate. Being hardy and easily transplanted, and comparatively free from disease, it readily takes rank among the few shrubs that should be considered indispensable on every lawn.

There are, we must always remember, certain shrubs which at first sight are not particularly interesting, and possibly a little coarse, but which, on further acquaintance, prove to blend picturesquely and harmoniously in combination with other shrubs on the lawn. The

DECIDUOUS SHRUBS

cephalanthus occidentalis is one of these shrubs, and we shall have occasion to present our respects to several other unpretentious species of this kind before we have finished discussing the merits of various plants suitable for the lawn. The amelanchiers deserve much consideration and respect for qualities of this unobtrusive but specially effective sort, and they have an added charm in their numerous snow-white flowers, blooming in mid-spring, and giving the plant a most interesting appearance. There is a Japanese species, amelanchier japonica, which is specially valuable for the picturesque way in which it masses its foliage. The witch hazel, or hamamelis, which is not a true hazel, is less valuable, and the same may be said of the picturesque halesia tetraptera, or silver bell. Celtis occidentalis, the nettle-tree, is another of these shrubs, or small trees, that have charm combined with simplicity and a blending quality. It has numerous slender branches and leaves, and a fine habit.

We desire to express our most profound respect for all these shrubs, not because they are more beautiful than other shrubs, for they are not, but for the simple reason that they sustain such fine relations with each other and all portions of the group, and lend just the touch of wildness needed to bring the entire effect into harmony and sympathy with nature. Quite different in effect, although more showy, are the masses of the rhus cotinus, purple fringe or smoke-tree, the flowers of which have a purple or misty appearance, relieved by the mingling with them of good-sized, rounded green leaves. Yet the rhus cotinus is somehow a coarse shrub, and requires to be planted with care and judgment, singly or in masses by itself, although its general effectiveness makes it impossible for us to neglect it in selecting lawn-planting material.

The white fringe (chionanthus virginica) has scarcely any relation to the smoke-tree except in name and vigor of growth, for they both make trees rather than shrubs. The foliage of the white fringe is light green and glossy, and the flowers are like the most delicate white lace, blooming in profusion, and giving the shrub, or tree, its chief value. It should stand alone, or on the outskirts of a group of trees or large shrubs.

Of the hibiscus syriacus, rose of Sharon or althea, it may be said that it has at least one attraction in its bloom in August; but otherwise it is stiff in habit, somewhat coarse in appearance, and exhibiting dull-red, brick-colored and bluish tints that are not pleasing to many.

Hydrangea paniculata grandiflora is at the present time one of the most popular shrubs used on the lawn, but in reality its chief value lies in its late-blooming quality, the large lumpy clusters of white flowers changing in fall to purple, red, and crimson, being showy rather than beautiful, and the foliage uninteresting and somewhat insignificant.

Just at the present time there is almost a craze, speaking horticulturally, for the California privet, which is really a Japanese species, ligustrum ovalifolium. There is no doubt its popularity is based on considerable foundation, for it transplants easily, grows rapidly, holds its leaves till winter, and makes a great solid mass of dark shining green, and has, what is a very unusual qualification in a shrub, the ability to grow tolerably well in the shade; and the word " tolerably " is used advisedly, because no shrub can be expected to do as well in the shade as in full sunlight, and very few will succeed at all well in the shade. The difficulty with the California privet is its tendency to shoot up into stiff, top-heavy forms and grow bare and leggy at the base. It fails, moreover, to present the variety of form of leaf and light and shadow and coloring that give so much pleasure in the hazel, amelanchiers, and several of the viburnums and dogwoods. There is a privet, also from Japan, called ibota, that has a rounded, attractive leaf, and a much better because more bushy habit, than California, or ovalifolium. For the general purposes of the lawn, the old, well-known common privet, ligustrum vulgare, has decided advantages when compared with the California, because, though less shining in leaf, the foliage is more spreading and thicker at the base of the plant.

It should hardly need to be said, although it does need saying badly in some quarters, that the practice of clipping the privet into formal hedges, flat or rounded at the top, and into gate-posts and other human or inhuman devices, is to be deprecated. If all the plants are clipped like those in a formal garden, there will be a unity of effect, whether we conceive the design to be successful or not; but to clip a hedge of privet in different forms, and then to cluster against it unclipped trees and shrubs, is hardly defensible.

When we pause before a favorite shrub, like lonicera fragrantissima, the fragrant bush honeysuckle, and attempt to analyze the feelings that predispose us in its favor, we find that there is a mingling of regard for the useful and the ornamental, as displayed by the plant on the lawn, and the better we know and the more we use shrubs, the more we will come to give increasing weight to the useful qualities.

There are shrubs, as we have seen, that look better standing by themselves, and one of the most notable of these is the dwarf flowering horse-chestnut, with picturesque foliage like that of the ordinary horse-chestnut, and spikes or racemes of white flowers that make a fine effect, rising above the broad-spreading mass of the leaves. For a reliable shrub, the masses of which mingle well in any group and bear fine, sweet-scented flowers and broad effective leaves, no plant deserves higher praise for hardiness, vigor, and beauty than the large sorts of philadelphus, or mock orange, among which should not be included the lumpy, yellow-leaved kind that, in the hands of the nurserymen, has had a certain ephemeral popularity.

Among the good all-around shrubs that can be counted on the two fingers, and that are always welcome on the lawn, we find the rhodotypus kerrioides, ranking high in excellence. There are more showy shrubs, doubtless, but for a refined, graceful habit and delicate green foliage combined with considerable vigor, for a medium-sized shrub, and adaptability to different soil and climate, it is difficult to surpass the rhodotypus. It has, moreover, an air of the American woods, although it is Japanese; and is admirable in combination with shrubs coming direct from American hedge-rows. Rubus odoratus, an American native shrub of similar character, has less finish and gracefulness, but it is, nevertheless, full of suggestions of woodland regions.

As a general rule, we are apt to find most of the spiræas somewhat weedy in appearance and lacking in the solid vigor and marked picturesqueness we have a right to expect in shrubs that undertake to occupy the rank of all-around species. Among the two or three kinds that we might, perhaps, see fit to allow to aspire to this rank is spiræa, or neillia, opulifolia, or the native ninebark. It is a little coarse, but its vigor is so great, and it has so much ability to grow in the shade and in all kinds of soil, and arrange itself in fine, picturesque masses which at the same time combine well with other shrubs, that we find ourselves coming very near to allowing it the rank, after all, of a good all-around shrub.

It needs scarcely be said again that the green forms should be selected and employed, and not the gold. Almost as good a shrub in its way is a dainty spiræa bearing masses of minute white flowers in early spring, and known by the somewhat formidable name, thunbergii; and spiræa van houttii, also, is another species which is very distinct, but small, and graceful and effective when it is planted in the foreground in groups by itself.

A very different shrub, on the other hand, is the Indian currant, or symphoricarpus vulgaris, which is less distinguished and refined in character than the spiræas just mentioned, but has a pleasant look of the woods, and a close, low, picturesque growth, suited to banks and other parts of shrub groups, where it serves to grade down successfully the larger shrubs to the herbaceous plants and grass.

An attractive shrub of the woods that is little appreciated is baccharis halimifolia. Its rounded masses of picturesque green foliage have a unique appearance which is, nevertheless, suggestive of naturalness and the forest.

Myrica cerifera, the candleberry, is another good shrub, of somewhat lower growth, that has never had proper appreciation given its small, picturesquely massed, dark-green leaves, which merit almost as much admiration as the azalea or rhododendron.

A distinguished family of shrubs, a family whose excellences the horticulturists should never tire of praising, is the viburnum, or snowball. First in general reputation comes the graceful and refined viburnum opulus sterilis; then the common, high bush cranberry, viburnum opulus, bearing fine fruit in autumn, and white flowers in June; and, finally, we find the most lauded kind in viburnum plicatum, the Japanese snowball, with dark-green, crinkled leaf, solid, erect form, and large snowball flowers, holding to the stems for weeks.

In spite, however, of these being the kinds that have thus far received general appreciation, it would hardly be going too far if there should be pointed out at least half a dozen other species of the family, any one of which has more excellent qualities for general use on the lawn than those, good as they undoubtedly are, which have just been mentioned as so popular, and, strange to say, the kinds that are referred to are little seen on the lawn in this country.

Most notable of these viburnums is sieboldii, or japonicum, with its great, massive, crinkled leaves, and large, red seed-vessels in August and September, and white flowers in June.

It grows into great specimens, fifteen or twenty feet high, the equal of which, among shrubs, in excellence, can hardly be found anywhere. Darker, and bearing finer fruit, viburnum lantana, while being, perhaps, less picturesque, has great value for its adaptability to combinations for general effect on the lawn. The pear-leaved viburnum, V. pyrifolium, is little behind the last in picturesqueness, and easily recognized, with its pear-like leaf, as a relative; and V. prunifolium, a specially hardy kind, growing in all soils, and richly colored in autumn, should certainly be classed as a shrub of the highest value. When we have added to these viburnums the best of all, perhaps, for vigor and adaptability, V. dentatum, we will find that the viburnum family forms a host in itself as a source of good lawn-planting material; indeed, we can readily imagine a lawn most amply furnished forth with plants selected from its ranks, for there are many other good species within the limits of what has been long known as the common snowball family.

There are a few other families that, like the viburnum, seem especially adapted to furnish almost all the lawn needs, in the way of shrubs and vines, from their own ranks, and among these few stand conspicuously the roses. Bushes large and small, climbers and creepers, are all found among the various species of roses, and whether we use bushes of dark-green, healthy foliage, bearing pink and white single flowers and great red seed-vessels, like the rosa rugosa; or delicate, picturesque masses, like those of R. laxa or R. multiflora; or R. rubiginosa, the American sweetbrier; or climbers, like the wild prairie queen, R. setigera, most vigorous and picturesque of all; or R. wichuriana, with dark, carpet-like masses of leaves studded with numerous white single flowers, we find a delicate, refined, but entirely healthy charm about the whole of them which is quite unique, and which, fortunately, perhaps, does not even suggest a comparison with any other family.

All these remarks are intended to apply to species of roses, and not to varieties, such as the mildewed and rose-bug attacked hybrid perpetuals, which, when they are required for their beautiful flowers, should be relegated to some secluded spot in the garden.

One distinguished family remains for discussion, although there are many more, did space permit, that we would find profit in considering. We refer to the hawthorn, distinguished for ages in English song and story. Unfortunately for the romantic associations that lead us to value a special plant which has become part of general history, we find that the English hawthorn does not thrive here, owing to a blight to which it is subject, although, when it does bloom, it easily keeps the high reputation it has so long held. Fortunately, when we are seeking for a good species of hawthorn, we are not obliged to pause at the English form and wonder whether we had better take our chances with it, for there are American species that are entirely healthy, hardy, and suited to all soils, and possessed of a richness of shining green color and a picturesqueness of layer-like masses that is altogether excellent, and worthy of the most distinguished place on the lawn. Cratægus crus-galli, the cockspur thorn, and C. coccinea, are two good species, and there are at least a dozen others.

EVERGREEN TREES

TO persons who have visited, and become familiar with, the country seats of Europe, it is difficult to explain why evergreens in America fail to thrive as well, and live as long, as they do over there, for the soil and climate are not very unlike; it may be that they are a little moister on the other side, perhaps, and less liable to sudden changes of temperature, but, on the whole, they are very similiar. One of the most difficult lessons for foreign horticulturists to learn is the necessity, if success is to be assured, of adapting the selection of plants, and their treatment, strictly to the results of experience in America, without regard to Old World standards. Whether an evergreen is likely to live long is not so much the question as whether there is a likelihood of its beauty exhibiting a reasonable amount of endurance. If the lower limbs of trees are likely to die, or a rusty appearance sets in, at various points, as a result of disease and attacks of insects and the vicissitudes of different seasons of heat and cold, we will hardly like to try them, although there may be a fair chance of their lasting in some shape for many years. Naturally, every one, especially if he is inexperienced, is tempted to try evergreens, particularly if they are beautiful, which may be all well enough if he does it with his eyes open; and, therefore, the author takes up the question of planting evergreens and comparing their excellence, to the end that experiments of the reader in this direction may be fraught with as little loss as possible.

It may be said of pines that many of them have both beauty and picturesqueness of trunk and foliage, even when exhibiting the last stages of decay, and their young, fresh growth can always be depended upon to be charming. When we speak of pines in America we are apt to refer to white pines, which, in our minds, represent the somewhat typical and, without doubt, the best conception of the general character of the family. It has been frequently said that the white pine, in America, performs relatively as important a part in the landscape as the palm in Central and Southern America, and above the value of mere visual beauty comes the delightful sound of the wind in the white pine, and the fresh smell and elastic touch underfoot of its needles.

In addition to the white pine, there are the Swiss stone pine (P. cembra), and the dwarf mugho pine, which possess dark beauty and healthy, long-life vigor. It would be not unfair to bunch together firs, spruces, retinosporas, American and Chinese arbor-vitæs, cryptomerias, Lawson's cypresses, sequoias, and junipers, as comparatively more or less unsatisfactory after they have obtained maturity.

Like all general statements, there are marked exceptions to be considered, and notably among these are the red cedar (juniperus virginiana) and the Oriental spruce, which are most picturesque, and fine at any age, although they are hard to transplant, unless set out when young, and with great care. This group of comparatively unsatisfactory evergreens might easily include the yews, for very few of them are really hardy, except the Japanese taxus cuspidata, and one or two dwarf American species. Yet the English yew, T. baccata, hardly ever dies from the accidents of heat and cold, although it occasionally browns in early spring, when it quickly recovers, and retains, with great persistency, its former picturesqueness. Another exception is found among the spruces, in the case of the common hemlock, abies canadensis, which lasts in good order a long time, and seldom suffers in winter. The trouble is, it is generally used in hedges, where its crowded condition tends to seriously impair its capacity for long life.

One of the chief drawbacks to the use of evergreens is met in the accidents that occur during the process of transplanting. Like most accidents, many of them, we are sure, might have been avoided by intelligent care, which means puddling the roots with mud, and keeping them absolutely from the air, until they are actually set in the ground. The small fibers of the roots of evergreens are so sensitive that they readily shrivel up and die when exposed directly to the adverse influences of the sun and wind.

EVERGREEN SHRUBS

THERE is a charm about evergreen shrubs that attracts every one; why, it is difficult to say. Perhaps it is a kind of artistic quality and dainty disposition of the leaves that is peculiar to the race, and which certainly does not pertain in the same degree to other inhabitants of the lawn. Then, moreover, it is a common fancy with people whose knowledge of plants is limited, that they must have evergreen leaves in large quantities on their lawn, in order to prolong its beauty throughout the winter, forgetting that a birch, or red-twigged dogwood, or a great, naked, sturdy oak will be able to easily enter into comparison with the finest evergreen shrub for the award of superiority in picturesqueness and abiding charm. Furthermore, we should remember that there are deciduous shrubs which in the American climate grow better and bloom more freely than they do abroad, and that are altogether quite as picturesque-looking as the evergreens, even in winter.

The reason that has induced the author to dwell thus at length on the comparative beauty of deciduous and evergreeen shrubs, is that we will be obliged to confess, as to usefulness, that evergreen shrubs are not always hardy under the stress of American seasons; certainly not as hardy as deciduous shrubs. There is no intention, on the author's part, to discourage the planting of evergreen shrubs. Far from it, for he would plant just as many as he could afford, with due respect to the appearance of the place; but he would do it always with the feeling that they should be set in sheltered places, and under the protection, but not shade, of neighboring trees, and he would feel proportionally proud if he succeeded in growing them, and not too much cast down if he did lose one now and then.

It would be only fair, after making such a broad statement about the tenderness of evergreen shrubs, to produce at once the apparently necessary exception to every rule, in the instance of the two special hollies of America and Japan, known respectively by the name ilex opaca and ilex crenata. It may be said, without fear of contradiction, that these hollies, when they are once successfully transplanted and vigorous, are entirely hardy, and in the case of the Japanese species the transplanting is easily effected. The American holly is shy in this respect, and needs coaxing, by transplanting at an early age, but, like its Japanese relative, since it does not come to maturity early, it retains its full beauty for a long period of time. If the reader wishes to realize what its beauty can become under favorable circumstances, he should visit some of the natural holly groves of New Jersey, and see, in March, when vegetation is all dormant, a number of hollies, twenty feet high, bearing large, picturesque, shining leaves, disposed in graceful masses, and further adorned with numerous large, bright-red berries, borne in full beauty from the past fall. It is hard to conceive of any sensible reason that should be sufficient to account for the lack of popularity of this shrub, unless it be its slow growth and shy transplanting.

But, fortunately, we are able to turn to our other holly, ilex crenata, and say that here is a shrub that is as easily transplanted, as hardy and rapid-growing (at least a foot a year, when once established), as some of the best deciduous shrubs—berberis thunbergii, for instance; and then the shining light-green leaves, like those of box trees, how beautifully they are disposed in picturesque, close-set masses, relieved from any suggestion of stiffness by the young growth that spreads like a halo around the foliage. Such plants as these hollies, as distinguished from deciduous shrubs, are unsuited for grouping with other plants, being really too precious to exhibit otherwise than in masses by themselves.

As we turn to the less hardy evergreen shrubs, and assume the attitude of accepting their small weaknesses, and of prizing success the more because they are sometimes weak, we find ourselves valuing azalea amœna as one of our choicest possessions, with its thick masses of small, bright-red flowers in spring, and its thick, rounded leaf contours, dark green in summer and bright red in autumn. Cratægus pyracantha, the evergreen thorn, has similar advantages in picturesqueness of foliage of a more irregular sort, which often browns in winter. The mahonia, or berberis aquifolia, is another picturesque evergreen shrub, exhibiting the most varied shapes and coloring of foliage, and though it browns often, it seldom actually dies, and, consequently, deserves consideration as a lawn plant.

It may be said, however, though the author is not prepared to allow that it is justly said, that in America the best hardy evergreen shrubs are the rhododendrons, and it must be certainly allowed that they are the most popular. It is far from the author's intention to combat this idea, for he, too, would have rhododendrons on every lawn, but it might be well to mention that rhododendrons have a way of dying badly if they are not born of the proper strain. Everestianum is an instance of this hardy strain, displaying fine foliage and purple flowers, with none of that vivid red that betokens a tender tropical breed. Yet there are good red kinds, like H. W. Sargent, but certain reds are, nevertheless, to be feared on account of this hectic flush, which seems to betoken a constitution unsuited to stand the strain of the hot sun and sudden cold winds of American springs. There is good reason to believe, reason based on long practical experience, that the rhododendron suffers in winter more because its young wood has not succeeded in ripening properly, being retarded in late summer by droughts, and pushed into new growths by rainy autumns. In view of this highly probable cause, it becomes easily evident that it is always a good plan to grow rhododendrons in strong, yellow loam, unmixed with peat, and in the open sunlight, where the new wood can healthily mature itself. It is no objection to this treatment, moreover, that it tends to develop greater numbers of the splendid flower clusters which are, perhaps, not to be surpassed for magnificence by any other bloom of the temperate zone.

The kalmia latifolia, or mountain laurel of American woods, is an evergreen shrub of the highest excellence, and although not as showy in bloom, it discovers to the observer who will take time to appreciate it a more daintily formed and exquisite flower than that of the rhododendron. The leaves are light green, and not close-set, but decidedly picturesque in their disposition, and while the kalmia equals the rhododendron in its ability to transplant in the spring without apparent check or injury, it should not be planted, any more than the rhododendron, later than August, on account of the bad effect of many winters on most all evergreen shrubs, and especially on those that have been planted too late for a good set of new roots to establish themselves during the autumn.

Among the many beautiful shrubs which come from the mountain ranges of Tennessee, Georgia, and North Carolina, there are few finer than the andromedas, notably A. floribunda and A. catesbii. The first is notable for the abundant white flowers it bears among the comparatively dwarf masses of dark-green, handsome leaves, and the second for the beauty of its large, shining green foliage, that is hardly surpassed by that of any other species of evergreen shrubs.

Every one is familiar with the light, pleasing green, close-set foliage of the tree box, and its contours are often fine, but we fancy its very familiarity in these days, when romantic association with childhood gardens and old colonial yards has again come into fashion, has lent it a slightly factitious value. It may be readily allowed that it is a picturesque shrub, easily transplanted, and valuable on account of its association, but when we come to compare it with the Japanese holly, ilex crenata, both for hardiness and variety and richness of charm, we will find it somewhat wanting.

HARDY HERBACEOUS PLANTS

IT is a little difficult to describe this class of plants to the ordinary reader, because they are both hardy and, in some cases, moderately long-lived. The leaves are not less beautiful than those of shrubs or trees, and the flowers are celebrated for their high range of quantity and quality; and, in fact, all that makes them at all distinct from shrubs is the peculiar habit they have of dying down every winter, and starting up again in the spring. When we turn to the woods, and proceed to gather wild flowers, we should remember that we are really gathering hardy herbaceous plants in most instances, for among them are included nearly all of the most attractive as well as most modest and exquisite blooms of the forest and field. In another place we have indicated their proper habitation on the lawn to be—and it will bear repeating—in the foreground of shrub borders, where they serve to round out and carry down the mass of foliage to the ground, and, also, we have found that they live and look well in nooks and corners, outside of the house, and at the foot of stone walls and fences.

It should not be inferred from anything that has been said here, that hardy herbaceous plants, or perennials, as they are often called, because they are hardy, have a mere weedy character that will afford excuse for neglect on account of their readiness to take care of themselves. It would seem to be an axiom in horticulture that there is no plant that is worth attempting to cultivate that will not repay the most liberal supply of nutriment and care of every kind. Naturally, herbaceous plants do not make an exception to this rule, any more than the common wild shrub of the fields, which is often the best hard-wooded plant we can use on our lawns. There is, consequently, something to be conceded to their forest-suggesting appearance, in their arrangement along the borders of shrubs, whereby an entirely irregular picturesque and wildwood effect will be produced. It is not intended to convey by this term " wildwood " the idea that there should be no carefully worked out design with reference to securing bloom in the different months of the season, and in grouping with relation to color and form, but only that by cunningly devised methods there shall appear to be a certain artlessness.

Consequently, it behooves us to see that no formal beds are designed that are unrelated to shrub borders, or corners of buildings and fences; and, furthermore, that the lines shall so blend with the border lines of shrubs that they shall practically merge into them. So much depends on the mass effect of herbaceous plants, as distinguished from their individual characteristics, that it is a good idea to set them thickly in the bed, with the expectation of, in four or five years, lifting and separating them, and adding or taking away material, and resetting them. It is only in this way that really satisfactory results can be obtained with herbaceous plants, for thinning out of thick planting must be attended to as well as the cultivation of the soil, whatever the kind of plant may be. Again, the wise horticulturist is he who, ignoring any question of expense, invariably secures the best lawn-planting material he can get, that can be readily transplanted and bought at a reasonable price, whether it be a tree or a herbaceous plant, for in the end it will be found to be the cheapest and best way, as the desired result can thus be obtained quickly, and the length of time the beauty of the plant can be enjoyed will greatly differ from that of the small, weakly specimen whose chances of life and vigor must be proportionally uncertain.

Lack of space will not permit us to dwell as long as might be desirable on the numerous beautiful perennials, but we will endeavor to look at some of the most attractive, and, at the same time, most useful and least weedy, members of the different families. If any one will take a catalogue and visit the several nurseries, and really examine the different merits and defects of herbaceous plants, he will soon be surprised to see how rapidly he is learning to shorten his list of kinds available for the lawn. It has been already intimated that a list of the good, all-around trees, and also shrubs, could be counted on the ten fingers, and the same result will occur if herbaceous plants are studied in the same spirit that seeks only satisfactory lawn-planting material, and not mere horticultural curiosities.

The same principle of treatment will lead farther, and will induce us to use large quantities or colonies of one herbaceous plant, and to keep in the shrub borders of the flower garden only the kinds that are not weedy looking, and that have sufficient solidity and symmetry of form to assimilate them more or less in general effect to the ordinary hard-wooded shrub. It will not take long to consider those strictly suited to the garden. Garden phloxes, as distinguished from annual phloxes, furnish a good illustration of such herbaceous plants. Their colors are rich, pure, and varied, and their stems clean and solid-looking, and of moderate growth. The same may be said of lilies, such as lilium auratum, the golden lily, white, spotted with maroon, and showing a wide gold band; L. candidum, the Madonna lily; L. longifolium, also a good white kind; L. harrisii (the Bermuda Easter lily), and the beautiful Japan lily, L. speciosum, both red and white; besides native field lilies, L. superbum and L. tigrinum. Lilies stand so firm and tall against a stone wall or solid mass of shrubbery, that we do not wonder at the praise the poets have given them.

Similar praise, except that they are not so tall, may be given the irises, with their beautiful, solid, simple leaves and remarkable flowers.

Iris koempferii, whose shades of purple, lavender, and blue, marked with bands of straw-color, appearing on flowers formed as curiously as those of any orchid, make it noteworthy in any association of plants, and only a little less distinguished than the German iris, the beautiful tints of whose different kinds last during a number of months of the year; finally, even the make-believe iris, pseudacorus, or yellow flag, shows a clean, firm finish of form that sets off well its yellow flowers, and renders it suited alike to the garden and the shrub border, or the edge of pools of water.

Single dahlias have a dignity of carriage, although spreading and picturesque in form and charming in single flowering effect, that seems to give them a place in the garden. There is a dainty, richly-colored, yellow-crimson flower of medium growth, the gaillardia, that has a form and color that suits it well to the garden. The erect forms and rich color of the gladioli also give their splendid flowers a right to take their place in the flower garden.

Of the lower-growing types, suited to formal places where they stand alone, as far as a background of trees and shrubs go, we find the brilliant, yellow, free, all summer blooming coreopsis lanceolata; and larger sized dielytra or dicentra spectabilis; and the sweet-william, dianthus barbatus, with flowers of various hues, red, white, etc., excepting blue and yellow, growing on solid stems a foot high, in flat terminal clusters; and the lovely carpet-forming, creeping, phlox subulata, with its broad patches, in spring, of minute red and white flowers.

A whole paragraph would not seem too much for the proper discussion of the merits of the narcissus family, the daintiest, perhaps, of all plants that belong distinctly to the garden. Narcissi, or daffodils, have long, narrow, simple, interesting leaves that keep erect and effective, and, above all, bear in a dignified way of their own yellow or white flowers half an inch across. Lovely sulphur and yellow colors specially characterize them, and the shape of the flower is curious and beautiful in every way, and quite difficult to describe. The poet's narcissus, the most beautiful of narcissi, snow-white at the base of the flower, with a crown of saffron yellow bordered with scarlet, belongs to the same section of plants. The best daffodils are N. pseudo-narcissus, the Lenten lily, with solitary flowers of a bright sulphur; N. bicolor, pale yellow, and N. princeps, white, and N. incomparabilis, with a larger flower and a shorter crown, two and a half inches wide, with a paler base.

The N. tazetta, or polyanthus narcissus, bears a number of fragrant flowers on a stem, and is white, with a crown of rich yellow. There are, in fact, scores of varieties of narcissi, nearly all of which are beautiful and early-blooming. The narcissi will succeed and look well in beds, but are, after all, dainty, retiring flowers, that would naturally seek retired places in borders of the garden or shrub group.

The humble but dainty and charming lilies-of-the-valley, and the cheerful yellow crocuses, both find suitable places in corners of the garden, or nooks of the shrub border. The anemone japonica is also a suitable and beautiful garden plant.

It may seem to the reader that the list of plants suitable for the garden is meager, and it must be confessed that it is, in view of the richness of the list of herbaceous plants which is rightly offered as beautiful in nurserymen's catalogues, but it will be found that their foliage, for the most part, is defective, and not suited for beds in the garden. It tumbles apart, or is weedy-looking, and makes an untidy appearance, especially towards fall.

There is, on the other hand, no difficulty in finding plenty of suitable places for all the many beautiful herbaceous plants we may desire, among and alongside the shrubs on the lawn. Here there will be a background that will relieve the uninteresting character of their foliage, and their weedy and irregular habit, especially toward fall, will be readily obscured in the general effect by the more dominant character of large trees and shrubs. It is, of course, evident that skill may be displayed in finding places for, and in selecting just the right quantities of, such plants as will render the scene brilliant and at the same time not uncouth and untidy.

The reader may have long admired such plants as hollyhocks and sunflowers, both in color and in form, but he may not fully realize, until he examines them again, that the foliage is shabby in summer, fading early, and the shape of the growth stiff and formal. And so it goes through the long list of spiræas, asters, asclepias, milkweeds, campanulas, dahlias, larkspurs, foxgloves, marsh and rose-mallows, poppies, peonies, rudbeckias, helenium autumnale, chrysanthemums, goldenrods, pyrethrum or chrysanthemum uliginosum—the giant daisies—and salvias.

Herbaceous spiræas are pretty and feathery in bloom, and should be distinguished from the hard-wooded kinds, like S. thunbergii. Their flowers are excellent, in many cases, for cutting and for fresh bloom, although when out of flower they are not specially attractive. The asters are a lovely family, familiar to many along roadsides of the Eastern and Middle States, the blue color of their masses making a decided and attractive feature in the landscape. Their foliage, however, is not especially effective in the garden, and the same may be said of the milkweeds, which are so fine in the fields, and also concerning the campanulas; but the dahlias should have a moment's longer consideration, because they bloom in late fall. The double dahlia has, unfortunately, a stiff, rosette-like flower, and needs staking, which makes it unattractive when out of flower. In form of foliage and flower, the single dahlia is much superior. As a fall flower, the chrysanthemum has great value, on account of the variety and great beauty of its many-shaped and tinted flowers. It should be remembered, however, that only a few kinds, of a simple button-like or lightly fringed shape, are entirely hardy, and on this account the lawn is a good deal debarred from using it.

Delphiniums (larkspurs) make us think of old gardens and childhood days, and their beauties are certainly great if the foliage would only make as good an effect in the garden as the flowers, which are dainty and charming, with unusual form and color. Rose-mallows and marsh-mallows have large, splendid flowers, both rose and white, but the leaves are not attractive in habit. Foxgloves are effective and curious in appearance, and poppies are splendid in red and scarlet color, but their foliage in the garden leaves much to be desired.

Peonies have splendid, large, early flowers, with pure rich tints of solid red or white, and develop into large clumps, which stand in borders and corners of the grounds, their leaves being gathered into loose masses which fall into decay in late summer. The single peonies, because the flowers are single and the foliage more compact, are better suited to the garden. There are a great many other attractive herbaceous plants which need planting by themselves, behind or mixed with shrubs and trees, where the peculiarity of their habit will not mar the general effect of the lawn, and where abundant bloom can be cut from them. Notable among these plants may be mentioned the great yellow sunflowers, six to eight feet high; the free-blooming, showy, black-eyed susans; rudbeckias; helenium autumnale; golden-rod or solidagos; the giant daisy, pyrethrum or chrysanthemum uliginosum, and the loose-headed salvias.

It seems to be treating so lovely a class of plants as the perennials with actual discourtesy to pass them by with no more detailed mention of the few named, and with entire neglect of the many beautiful kinds that are not named; but it has been the author's special object, throughout this book, to deal with the practical principles of ornamenting the home grounds rather than with the discussion of attractions of individual plants, and to therefore cite plants as illustrations of points which he may desire to make, and not as objects to be dwelt on with simple delight and appreciation.

AQUATIC PLANTS

THE growth of aquatic plants, although long performed in the tanks of greenhouses, is a comparatively recent acquisition on the lawn, where it serves, when arranged properly, to greatly enhance the charms of pools and streams of water and their shores, as well as the surface of fountain basins, on both public and private grounds.

A half cask filled with a little good mold and water may easily serve to secure for the smallest village yard the enjoyment of the charms, and they are many, of hardy aquatic plants. The shores of pools or streams of very humble dimensions may present a creditable exhibition of aquatic plants by accumulating, for their growth, deep rich mold along their margins and at their bottoms, although care should be taken to confine the roots of such kinds as nelumbium speciosum by means of boxes or bricks or stone partitions, to prevent an overgrowth which will soon occupy the whole surface of the water, to the great detriment of the picturesque beauty of the scene. The great leaves of the lotus are extremely decorative, and striking in effect, to be compared only—if we may be allowed so humble a comparison—to the splendid leaves of a pumpkin vine; and, like

AQUATIC PLANTS

some great rose, the flower rises on a long stem from the general mass, with seed-vessels borne later on that are strangely like the spray of a watering-pot, whence the botanical name, nelumbium.

The lotus of the Nile, nelumbium speciosum (and there is a yellow American species, luteum), is, of course, a name to conjure with in dealing with the ordinary reading public, but the beauty of several of the hardy white nymphæas, lying in picturesque clusters on the water, form, to the author's mind, far more attractive objects than the lotus on the pools and streams of the ordinary lawn. They seem always neat and finished in their design, and elegant and decorative beyond the capacity of words to express. The way the perfect white flowers are arranged on the shining clusters of green leaves, as they float on the water, is a sight to see for one's self, and not to read of in a book. There is a little white lily, N. pygmea, about the size of a silver half-dollar, that illustrates the truth of this statement with special effectiveness.

Unfortunately, there are only a few hardy kinds of nymphæas, or water-lilies, chief among which are N. odorata, the Cape Cod lily, and its varieties. There is a large white lily, N. alba candidissima, which is much used, and by the employment of a little shelter and heat, in the form of a tank of warm water, in even the smallest greenhouse, other beautiful forms of water-lilies may be secured for the ordinary lawn. Among the best kinds of these half-hardy water-lilies may be mentioned nymphæa devoniensis, nymphæa zanzibarensis azurea and rosea, nymphæa flava, and nymphæa sturteventii.

The capacity of the pool for beautifying its surface with aquatic plants does not end, by any means, with lotuses and water-lilies, for there are curious water-hyacinths of the pontederia family—a floating plant, with swollen leaf stalks and blue flowers, looking like an orchid; and the water-poppy, limnocharis humboldtii, with round, bright-green floating leaves, and solitary large sulphur flowers with three petals; and, finally, the water-hawthorn, with flowers in white, short spikes, which should be treated like the water-poppies. Then there is the water-anemone, ranunculus aquatica; and pitcher-plants, sarracenia purpurea; and arrow-heads, sagittaria sagittifolia; and the floating water-plantain, with white flowers borne on long-stalked, small, elliptical leaves; and on a little drier ground, irises, and one or two orchids, such as cypripediums; and, finally, the excellent yellow flag, iris pseudacorus, already mentioned.

In closing these few, and quite inadequate, remarks on aquatic plants, the author desires to point out that natural and quiet pools or streams seem to suit them best, although a fountain basin is well fitted for the same purpose, where a small drip of water or moderate spray disturbs but little the surface. Rock-bordered cemented pools for aquatics hardly ever look quite successful, whereas a drink-hole for cows in the meadow may support pond-lilies for years with the most charming and appropriate effect.

HARDY VINES AND CLIMBERS

NOTHING in the way of lawn-planting material probably contributes as much to the natural and picturesque effect of the home grounds as the climbing and creeping vines that may be used on it. It is not true, perhaps, as a matter of fact, but there seems to be nothing so wild in vegetation, so essentially natural-looking, as certain climbing or creeping vines. It cannot be said that all climbers have exactly this effect on one, for the ampelopsis veitchii or tricuspidata has, at maturity, a large, shining, elegantly shaped foliage, which attaches itself to stone walls like English ivy, and masses together like shingles on a house, one leaf over the other, in a dignified and civilized manner, as far removed as possible from the wild, tangling habit of the woodbine or Virginia silk. One of these well-civilized vines is the jackmanii type of the large-flowered clematis, with its delicate masses of large, star-shaped purple flowers, and with it may be classed the white henryii and lavender lanuginosa; but Japanese clematis paniculata is literally a wild thing, throwing out almost limitless quantities of small white flowers during the latter part of summer, and climbing over everything in an inextricable tangle. Of the same general wild character, in this respect, are clematis virginiana and C. flammula, while little behind them come the honeysuckles, which, though incapable of climbing to comparatively high altitudes as rapidly and in as effective a manner, have, nevertheless, unrivalled aptitude for covering stone walls and the borders of shrub groups in the thickest and most picturesque way, besides having the faculty of living and growing almost everywhere.

Somewhat less rampant and wild-looking, we find the common Virginia creeper, ampelopsis quinquefolia, performing the most efficient work of all, in covering with rapidity and perfection old stone walls, banks, and any large objects that need herbaceous covering, and, withal, presenting specially rich autumn coloring. Even more decorative than the ampelopsis are the different hardy grape vines. The native kinds, vitis labrusca and vitis cordifolia, grow rapidly, and are very effective in the way their leaves climb over stumps, but the Japanese vine, vitis coignetiæ, seems to have equal vigor and a magnificent and varied coloring of leaf in autumn. For training along the eaves of houses, and piling up in highly picturesque masses on the roofs of arbors, nothing in the way of vines can surpass the wistaria. It is a hardy climber, doing as well in crowded cities as elsewhere, and is apt to grow to a comparatively great age. There are several American species, but the Chinese purple is the best known, although there are several Japanese varieties, white, purple, and large-clustered, that seem to be nearly as good as any on the list. The wistaria is a little difficult to transplant and establish, but in a year or two it generally starts off to grow in the most vigorous manner.

We must not forget the trumpet-creeper, bignonia, which clings readily to rough walls and surfaces, and makes a fine effect in midsummer, with its great clusters of scarlet flowers and vigorous foliage, that piles up in fine fashion over rough stone structures, boulders, and old stumps of trees.

The actinidia polygama is another Japanese vine, bearing large, glossy leaves, and piling over walls and rocks in a decidedly rampant manner.

In the interest of decorative purposes we must not forget the Dutchman's pipe, aristolochia sypho, the large, light-green leaves of which climb up the wires on a porch, and lie over each other in a picturesque way. It grows a little slowly at first, but eventually reaches a great height, when properly trained. Dolichos japonicus is another vine that bears very large leaves, and reaches great heights with much rapidity.

Of the English ivy it would be well to say one word, not because any words are necessary at this late day to celebrate its rich beauty on stone work, but because, though most people know that in the Middle States it will not succeed on the south side of a house, and sometimes gets browned by winter on the north, its great value as a carpet under shrubs should be better recognized. In reality, the English ivy is a hardy plant, except when exposed on the south side of buildings, and will do well on the ground.

But after admiring all these vines, we turn to the climbing roses with a feeling that we will find there a quality of excellence that is not exactly equalled by any other climber. One of these roses is the wild setigera, parent of the well-known climber, queen of the prairies, and for vigor, clean healthy foliage, and profusion of single pink and white-striped flowers, combined with all the native grace of the family, it will puzzle the connoisseur to find any vine that will surpass it. The other rose is wichuriana, from Japan, which makes, on the other hand, a thick, carpet-like, spreading mass of small, dark-green leaves, and seizes quickly on the surrounding ground, studding the surface of its foliage with a profusion of small white and yellow-stemmed flowers. There is hardly any climber that will so successfully and beautifully cover the surface of an outside brick chimney or smooth stone wall when it is properly trained, and some of its hybrids promise to be still more effective

BEDDING PLANTS.

ALL over the country, alike on elaborate lawns and in poor men's door-yards, bedding plants, as represented by arrangements of coleuses and geraniums, are evidently popular to a remarkable degree. The writer desires to express his full appreciation of their brilliant attractions, and their value in landscape gardening schemes when they are properly employed. Nothing is more splendid and rich-looking, and thoroughly decorative, than an arrangement of cannas, coleuses, and alternantheras, when they are artistically combined; but, on the other hand, no plant effect on the lawn can be made more crude, garish, and vulgar than a badly designed and located arrangement of the same plants. It should be remembered that the more brilliant and striking a plant is, the more difficult it is to use it in such a way as to perform a harmonious part in the general scheme of arrangement on the smallest lawn; in truth, it proves, in practice, that the smaller the lawn, the easier it is to create an unpleasant, jarring effect with bedding plants. There are, of course, no plants that can be used carelessly, and in a crude and improperly related way, without due consideration for the other possibilities for beauty that the place may have, but so preëminently is this the case with bedding plants that, unless one is sure of the value of his scheme of arranging them, far better would it be to let them alone, especially in view of the fact that their cost is relatively somewhat expensive, as they have to be planted over again each year.

Bedding plants have, it is evident, a definite part to perform in the adornment of the lawn and door-yard, but their proper place should be strictly defined and adhered to, and that, it will be found, should be to act as a part of some more or less formal arrangement, such as may be made adjoining, or bordering, a building, or in a terrace or courtyard, or, best of all, in a formal flower-garden. In schemes of this kind, the peculiar brilliancy of bedding plants can be more readily managed than elsewhere. If such a scheme grows out of and definitely relates itself to the architectural scheme of the place, it is generally satisfactory, and, shut off in a garden, there is, of course, every opportunity for special effects, with little danger of injuring the general appearance of the lawn by showy formality. It is so easy to overload any spot with this brilliant, gorgeous form of plant ornamentation, that great self-restraint is generally in order when schemes for its employment are under consideration.

But when we have finally secured the best location for our scheme of bedding, and properly defined the extent of ground it is to occupy, we have still to solve the difficult problem of how to combine the different parts of each special scheme, or bed of color and foliage plants. It is at this point we meet the most flagrant failures in designs of this kind. The fact of the existence of general principles, that apply alike to all kinds of planting effects, seems to be entirely lost sight of. That there are such things as open pieces of low planting, and masses of higher planting, clustered around single points of tallest effect—in a word, that the sky line must be considered—is generally understood when applied to an ordinary landscape; but how few think of applying the same principle of arrangement to a small cluster of even a single flower-bed of moderate size. And yet all landscape gardening, to be good art, should deal with principles, and the principles should operate in every phase of the work, whether it be in a symphony of grass, herbaceous plants, shrubs and trees, or in one of alternantheras, coleuses, geraniums, acalyphas, cannas, or musas.

Let us consider for a moment a good example. It might consist, for instance, of a grass bank in front of a house, and be bounded on all other sides by walks, or a great gravel space. Massed back against the house would come great clusters of cannas, like small trees, and these different kinds of cannas would have a rounded outline, a sky line as it were, that would be simple and easy and graceful, and at the same time points and clusters of them would run forward among broad plantations of geraniums, with their sky line further accentuated by the presence of higher color, and growth of brilliant red acalyphas, the leaves of which are smaller than the cannas and larger than the geraniums. Outside of the geraniums would come irregular borders of yellow featherfews, nasturtiums, or the charming alternantheras, which grow only a little higher than the shorn grass is allowed to get; and out two or three feet into the general expanse of the turf little clusters of the rich-looking acalyphas would be allowed to wander, in order to prolong the effect of some of the chords of color that give special value to the appearance of the general mass. Naturally, it would depend on the skill of the designer, or composer, whether such a combination of the colors and forms of bedding plants would result in an artistic symphony or not, but there can be no question that the possibilities of fine effect would be great, far more than enough to tempt one to undertake a careful study of the existing environment and materials.

In the ordinary flower-garden, effects like these are entirely in order; indeed, more freedom of design may be displayed here than in most places, as the environment is simpler, and less complicated by surrounding scenery. If people would come to feel that badly placed and designed foliage-beds were as inartistic and as bad in their way as bad paintings, we would soon have more bedding effects that would be altogether admirable.

There remains only a few words to be said concerning the general quality of materials that can be effectively and satisfactorily used in foliage or color beds. It seems to be a fact, to begin with, that there are a great many plants often recommended for use in such beds that do not, and cannot, secure the brilliant and defined splendor of color and form that should characterize all designs of this kind.

The writer will not attempt to mention, much less describe, all the different kinds of plants that are suitable for bedding, but rather to point out certain qualities of a few notable species that mark their fitness for the purpose, and illustrate the special range of quality bedding materials should possess. Probably the geranium is the most popular bedding plant in this country, for it has a picturesque form, finely shaded leaf, and brilliant-looking flowers, although too much stress should not be laid on the flowers of any bedding plant, however beautiful they may be, because the leaf effect is valuable and satisfactory throughout the season, while the flowers fade in a few days, not to return, or if they return they bloom in a scattering or intermittent way. It would not be saying too much to declare that leaves should be the first and most important consideration, and not the flowers, and, above all, that plants exclusively valuable for their flowers, like lilies, should not be mixed throughout the mass of any bed, to the obscuration of the clearly defined design, for while there may be, and generally should be, a blending of colors and forms, there should not be confusion of effect. Clearly defined and striking colors and forms, should be sought in landscape gardening on the smallest and most formal plan of bedding, for bedding belongs in the foreground of the picture, and therefore rich color is just as much in place here as the green foliage, with its mysteries of effect and soft blending, is in sympathy with the middle distance and background of the picture. In bedding, we welcome all the color and glow we can get, while in the tree and shrub mass we deprecate the use of any other color than green, unless it be used with unusual self-restraint and skill.

Nothing more perfect of its kind offers itself than, perhaps, the different varieties of alternantheras, which are glowing and rich with color, and yet have a grassy habit that fits them better than almost any other plant to take the place of grass in all schemes of bedding. Next to this grassy effect, we have the yellow featherfew, and the nasturtium of charming habit, and the begonia, and after that, taking the place of the shrub in the ordinary landscape scheme, comes the geranium, already mentioned, and the coleus, most gorgeous and many-colored of plants. Coleuses and geraniums look well adjoining each other, while in the same relation may come bouvardias and salvias, for a mingling of some green color in the brilliant mass is often agreeable, when well managed. One of the best plants for brilliant effect in any bedding group is the acalypha, the brilliant red or reddish-green leaves of which take the place of some specially tall shrub, and make an effective key-point of color in the general mass.

It is, as already intimated, important to avoid all flatness or monotony of sky-line, if this unusual though entirely appropriate use of the word may be permitted when discussing foliage-beds. In this way cannas and banana plants, and other tall and large-leaved species, perform an excellent office in the bed, where they need not necessarily be given the middle or single point of effect, but may do excellent work in three or four places, or even entirely in the background, as when a bed is made along the side of a building. Like the shrub group, there need be no formalism or monotony in design, but these tree-like effects of cannas and musas, or bananas, may be arranged on a clearly defined and well-studied plan that will produce the utmost variety of color and form and outline, and even a symmetry of its own that is entirely harmonious and fitted to the ground and environment, and yet not in any way set or stiff.

POOLS AND STREAMS

THERE is something so delightful in the living, moving presence of water in landscape that it is no wonder that the prospective purchasers of places look for the visible existence of a bit of sea or lake, or even a pool or stream of water, which is either actually or potentially effective. The variety and characteristic charm of the sea, lake, or river is as valuable and genuine a possession, when you have it before you, as the sky, and, in its way, just as beautiful and precious; but when it comes to the little pool or stream that you will have on your land if you buy it, or that you may have if you succeed in translating potentialities into realities, the problem presents the usual considerable difficulties. There is nothing so deceptive, and difficult to estimate beforehand, as the possibilities of water as a harmonious and typical feature of a country place, because water is such an elusive feature. And it is especially so when the area of its display is limited within the small confines of an ordinary country place. It has a way of overflowing its banks, of drying up in a day, of disappearing in the earth without warning, of growing muddy, or green with scum—of doing eccentric, unaccountable, and unavoidable things that will be the despair of those unfortunate beings who may be, or may feel themselves to be, responsible for them.

If you must have water on your place, therefore—and why should you not, when its presence, under successful conditions, gives a delight and solace that would be worth several considerable failures sustained in the effort of obtaining it?—it is always wise to take a middle and conservative course, using pools and streams only when you can discover satisfactory evidences of the existence of unfailing sources of supply from living springs, or from some other large body of water that is practically inexhaustible.

The author would not like to attempt to discourage the would-be contriver of water effects from digging pools and streams on his place, and striving to make the most charming picturesque and perhaps natural effects. It is his right, if he is a person possessed of a vivid imagination and sufficient means, to seek to realize what to him will be one of the attractions of his place; and there is not the slightest doubt that he may succeed in constructing a bit of water that will look natural and never dry up, and never look foul, and, above all, fit in as a perfectly satisfactory element of the landscape picture. The use of the word "may," however, is done advisedly, and under the deep sense of the mutability of human effort when put forth to accomplish the difficult task of introducing pools and streams of water artificially as a pictorial effect on home grounds. Rather would the author, though perhaps he is too timid, prefer to study the bits of water he already has, and try, if he thinks it feasible, to develop the beauties the possibilities of which are plainly evident, than to go farther afield and lose himself in the bewildering mazes that beset the path of the imaginative creator of sheets of water, all fresh and made in accordance with the latest ideas upon the subject of constructing artificial pools and streams.

The first step in the development of the bodies of water we have, into those we would have if all the circumstances favored us, is the preparation of the shores of the water by deepening them, removing āll mud and débris, and preparing, at certain points, masses of good soil for the reception of any plants we may set out. In reinforcing these shores, it is well, without making petty, small curves, to vary their outlines to a considerable degree, by deepening what would correspond with bays on a larger scale, and building out the points, which are sometimes improved by the presence of a rock, and in artificially lifting up the bank with soil, while the hollows of the indentations of the edges may, at some wider opening, present a pebbly or sandy beach.

The introduction of artificial islands in a pond or lake is apt to be a dangerous experiment that, even if it succeeds in assuming shape harmoniously, and in no petty and undignified manner, will be somewhat difficult to retain uninjured under the stress of storms and the action of water; and here, again, we find it necessary to exercise restraint in the use of rocks, unless they abound in the immediate neighborhood, or unless some bank needs artificial support, the necessity for which seldom, indeed, arises. The proper treatment of brooks is similiar to that which is suitable for pools, the deepening and widening at certain points bearing a fixed relation to the supply of water and the general normal size of the stream.

If the reader will imagine a hillside, or sloping lawn,

extending down to a small lake or pond or running stream, we can try to see what ideas will suggest themselves as to different ways of treating the water. In the first place, it should be established that it is not well to attempt to make artificial pools and streams of water. It can, of course, be done, but it is generally difficult, on account of an imperfect supply of water, or else of a lack of fitness of the conformation of the ground for the water effect that it is possible to obtain. Flooded territories are not apt to be satisfactory in appearance. The conditions do not seem to be normal, the landscape not quite natural-looking.

The treatment of the natural stream or pool is a simple one in itself, and it is hardly possible to do anything but widen or narrow it, so as to create some specially fitting effect, or to plant it to a limited extent with water-lilies and aquatics, always being careful to arrange for large spaces of open water. There are also agreeable relations to be established with territory adjoining the water, and these afford much opportunity for study.

The site of the house should be so placed, whenever possible, as to get the best views or glimpses of the pools or streams, and the walks should also be laid out so as to gracefully and naturally pass out and in the masses of shrubbery or groves of trees, so that from the house beautiful vistas ending in water landscape should greet the eye. When the water can be secured as an important or dominant feature of the landscape, the arrangement of even the smallest pools can be so contrived as to enhance the charm of the water element in the general scheme. Though the place be so narrow that it is only possible to carry a straight walk from the

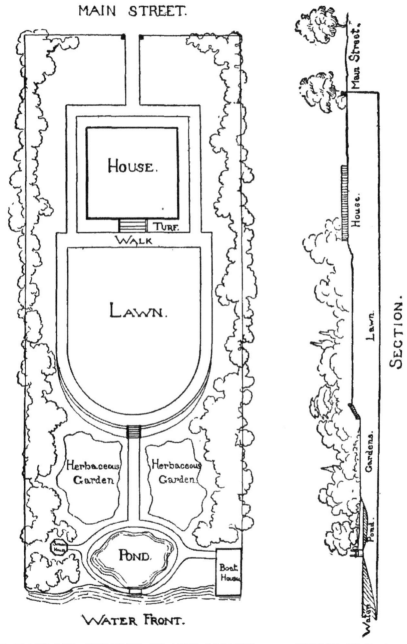

VILLAGE LOT, ONE-HALF AN ACRE, LOCATED ON STREAM

SECTION OF SAME

house to the water's edge, yet either boundary of the place may be shut in with a solid mass of trees and shrubs, and in front of them could come, with excellent effect, a garden of herbaceous plants, blooming throughout the different seasons of the year, making the plants more or less suited to the low ground gradually predominate as the shore is approached. Near the shore, the

SUMMER-HOUSE ON WATER

straight path could be forked out, enclosing a pool, or miniature landlocked bay, which might be scooped out from the stream, and dotted with pond-lilies. On either side should be natural shore, for it is a pity to use artificial-looking cut stone on the shores of any stream; the water is so much more attractive lapping up on a pebble or sandy beach, or even a border of grasses and rushes.

Excellent effects may be produced by making a little

eminence of earth, if it does not already exist, on one side of the lot, and erecting thereon some rustic summer-house, or vine-covered pergola, commanding a view over even a small reach of stream or lake. Down by the other corner of the lot may stand a little unobtrusive boat or bath-house, also vine-covered. (See illustration on page 155). None of these buildings need be made obtrusive-looking, but, in every case, special consideration could be given to varying and renewing, in some other form, the æsthetic value of the water in the landscape, just as we found an agreeable effect by a little pool at the foot of the land, the sight of which seemed to carry the eye with more pleasant feelings of expectation to the larger spaces of water beyond. All sorts of variations of the line of the shore and the height of the contour of its immediate borders may be effectively devised, and by means of rushes, cat-tails, pond-lilies, and other aquatic plants, an effective and beautifully picturesque foreground be given to the water itself. The imagination may revel in many plans for such water effects in connection with the home, but in undertaking to carry them into actual effect, one will find it always needful to remember that the house must be considered with relation to open space in front, or back, or around it, and that the shape of the ground, and any natural charm it may have originally possessed, should not be forgotten in the desire to accomplish some radical scheme of improvement.

WOODLANDS

THOUGHTS of grateful shade, and flickering bars of sunlight falling across the varied and dainty leafage of woodlands, have always filled the soul with delight. The more natural the forest glade, the pleasanter becomes a stroll through its leafy aisles, and the masses of its trees and shrubs and vines have a wonderful power of refreshment and restfulness. It may be hard to believe, but it is a fact nevertheless, that it is almost impossible to create by human skill a perfect imitation of a natural forest glade, for there are thousands of little touches of genuine wild beauty that no horticulturist will be able to produce, and it seems to be essential to the highest enjoyment of genuine woodland effects for every tree and shrub to have grown spontaneously, and to have been left by itself. This principle of leaving well alone in the smallest wood lot of actual village home grounds should, therefore, guide us in all our dealings with large or small tracts of woodland, for any extra planting or pruning, beyond cutting out dead limbs, or an occasional tree that is growing too close to its neighbor, is sure to impair the special charm of such places, which seems so easy to obtain and is so difficult to hold uninjured.

Every care should be taken of the trees by destroying insects, and renewing the top soil that has been washed away from the roots, but it is, as already stated, useless to attempt to plant fresh trees and shrubs; firstly, because they would not properly thrive in the shade and among the tangled roots; and, secondly, because the natural effect of the scene would be destroyed at once.

In the same way, wild flowers, or what are termed by the horticulturist hardy herbaceous plants, find it difficult to look exactly happy, even when they are skillfully planted inside of masses of woodland, with the idea of making them look natural. A little of this natural effect may be accomplished by clustering small colonies of particularly wild-looking flowers, like asters and daisies, on the extreme borders of woodland, where the sunlight can reach them, for it is an axiom, scarcely needing repetition, that no plant (there are a few striking exceptions) does as well in the shade as in the sunlight. When you find flowers blooming in the shade of forest glades, you will also find that it is very much a case of the survival of the fittest, under difficult conditions, wherein many die in order that a few may live.

On the other hand, although woods should remain natural, they need not look untidy or forlorn, or as if they were lapsing into decay, yet neither should they look as if they had been swept and garnished. The removal of a dead stump or a fallen limb, or any positive rubbish, is, of course, always in order, but to rake up leaves in actual woodland indicates ignorance and lack of regard for the trees when attempt is made to deprive them of the protective and mulching value of the fallen leaves; and, moreover, no other possible carpet, whether of grass or bare earth, can equal the pleasing effect of the coloring of the brown leaves as an accompaniment of the lights and shadows of the forest.

There are, however, many other ways by which the health of woodlands can be improved from year to year, for around the roots of trees in bare, unfertile places, from which the leaves have blown, good soil, mixed with manure, may be applied several inches thick with advantage.

Then the streams that flow, at times, through bits of forest can be cleaned out, and retained in their own channels, thus relieving the trees from damp, and maleficent influences.

So far, we have considered only things such as will make for the beauty and health of the woodlands; but when we come to foot-paths, we shall encounter features that, though necessary, can offer no reason for existing except their necessity, and should, therefore, in the interest of the woodlands, be limited to wandering trails, just wide enough for two to walk abreast, and reaching around large portions of the territory without crisscrossing them with a network of foot-ways.

Cultivation, in the form of digging and plowing, seems always to be out of place in woodlands, because it tends to disturb the mulching and enriching process that is being fostered by the shade and by the fall of leaves; and so, likewise, the paths or trails should have little trimming, only just a rough levelling to prevent water from settling in pools, and there may be added, with advantage, an inch or two of sand or gravel, to kill weeds and render footing more dry and agreeable.

It is mischievous negligence that allows cattle to wander in the borders of woods, destroying the lower foliage of the trees, and with it that beautiful effect of the tree growth when it is encouraged to droop down to meet the low shrubs and long grass of the meadow. It is on the edges of these woodlands bordering on meadows that one finds the most charming parklike effects, the true ideal type of pastoral designs, where Nature has been just enough influenced by the hand of man to give her the human interest that should be associated with all attempts of the landscape gardener. A bit of an old fence, a log, a ditch, gives a living sentiment to the picture, of which one feels the need in any park or home grounds. Pure, unadulterated Nature is all very well for mere sentiment, but an old lane, with its cows and sheep-dog, and hedgerow of dogwood, pepperidge, and liquid amber, and wild clematis, clematis virginica, will satisfy the spirit more days in the year than any Rocky Mountain glade, if we could reach it, where the foot of white man has never trod.

In order to explain better the bearing of these remarks, the writer desires to call attention to the attempts that are often made to imitate Nature by planting thick, tangled masses of native plants, so that, every way one looks, one is attracted by broad stretches of trees and shrubs that for a moment seem to have sprung up spontaneously where they grow. We gaze at these tangled masses of trees and vines, and are, for a moment, delighted at what we believe to be veritable Nature; but unfortunately, we are always bound, sooner or later, to be undeceived, and then, under the influence of the disappointment of deceit, we begin to look about us, and then to find that it is, after all, but a bit of theatrical effect imposed on the general landscape, with which it does not in reality harmonize, and with which no human hand can make it harmonize, because the remainder of the scene cannot be all wild. Moreover, though we exercise the deepest cunning of the art, and maintain it with the greatest care, the marks of the cultivator on the attempted wild plantation will be soon seen and noted, and not always felt to be satisfactory by common, sensible folks.

It is for these reasons that the writer deprecates all interference with masses of woodlands, except to foster them and keep them growing in the most satisfactory way, and to leave them absolutely untouched by horticultural skill except that which undertakes to remove excrescences, superfluous branches, dead wood, and clogging pools of water, and to allow not even maples and chestnuts to be planted along edges of woods, for fear they will injure the native woodland effect, which it is so easy to lose and so hard to restore. Setting aside all mere sentiment, and looking at the subject of imitating woodland effects in the sensible way which is always, as we have before said, the most artistic way, we will feel, as we look out on a well-designed park or home-ground treatment, that it is good to see great groves of trees intermingled with shrubs, and vines that cluster over walls and fences, and are arranged and cultivated so that they will grow and thrive for a number of years with little pruning, transplanting, or other change. They can be managed so as to make us think of the most charming effects of woodland and meadow, and yet not for a moment deceive us, but make us, instead, exclaim how well the grouping is contrived for the open meadows and lawns and long vistas of the place, and at the same time for the individual exhibition of the native charms of the trees and shrubs. The buildings, fences, and other structures necessary for human comfort and convenience will then preserve an agreeable relation to the plantations, and not jar on our sensibilities, as they must in the midst of the most perfect attempt at an imitation of Nature. In this way woodlands, untouched except for actual maintenance, fall into their proper place as one, and only one, of the most desirable features of home grounds or parks, and need not be insulted by futile attempts to imitate, and even improve, their charms, by planting other trees in their midst, after the method of certain horticulturists who fail to comprehend their proper function. To show how these woodlands should be allowed to furnish us with types and suggestions, rather than to be interfered with by so-called improvement, it is well to remember that the very oaks, maples, chestnuts, tulips, lindens, and ashes that constitute these woodlands, and with which we should find it so difficult to create a genuine bit of woodland, are found of inestimable value in making park pastoral effects of masses of trees and shrubs, and single, slightly detached specimens.

It is again evident that, in designing the arrangement of home grounds and parks, our first duty is to frankly preserve, without attempting to imitate them, the existing beauties of the place—woodlands, single trees, rocks, knolls, and meadows—and to only add such arrangements of trees and shrubs and flowers as will enhance and perfect the special charms that are native to the place, and at the same time not interfere with the comfort and convenience of the people who live there.

THE USE OF ROCKS

THE charm of rocks in a landscape is so evidently valuable that there is a natural desire to use them in lawns and gardens. But experience teaches us that their use is fraught with much difficulty; and for the very reason that rocks are such positive-looking objects, they should be grouped and massed with the utmost taste and discretion, otherwise the result is sure to mar the landscape, and in so disagreeable a way as could not seem possible at first sight. And here, as elsewhere, we find again that existing facts and conditions must be allowed to inspire us. Surrounding territory may not show the slightest suggestion of outcropping rock or loose boulders, and in such cases we need surely feel no desire to introduce a foreign element in the scene which can hardly fail to make it discordant. One rock in the wrong place, looking as if it had fallen unawares out of a cart, is totally out of place, while a number of rocks, spotted about in quite promiscuous fashion at the base of a slope abounding in large and small masses of stone, will often help the beauty of the landscape, especially if a footpath is present that appears to be here and there deflected by the presence of the rock.

In the case of rocks in landscape, it is very much the same way as with the man of ten talents in scripture. To the territory abounding with boulders and out-cropping stone, more will be brought, and in the region where there is a dearth of stone, even those that happen to be there will be removed. Thus, for example, take a trench, or stony ditch, through which a natural stream runs, and you wish to make it more picturesque, and enhance its possibilities as a characteristic and beautiful feature: even if it be cut through solid rock you may add still other rocks and boulders, and, setting them here and there, secure the finest effect possible.

On the other hand, if the cut happens to go through clear, solid earth, with hardly a stone, you will be very careful to remove any accidental-looking stone that may be seen. The writer does not attempt to lay down absolute directions, covering any and all cases—that is impossible—but he merely indicates the way in which the trained mind would, in such matters, set to work.

In the same spirit he wishes to emphasize the fact that the use of rocks should always be taken up with the utmost deliberation, because the misplacement of such objects as rocks is much more serious than the improper location of plants. The plants have, as a rule, a beauty of their own, irrespective of position, which is much greater than that of individual rocks, and, besides, plants have a certain indeterminateness of outline that is always less offensive under adverse circumstances than the rigid and sharply outlined masses of rocks.

What the writer especially wishes to explain is, that rock-work should have a definite fundamental scheme, springing naturally from the shape of the ground and nature of its rocky contents, and that the carrying out

TREATMENT OF ROCKS FOR STREAM AND BRIDGE

of this scheme had best be done in a large and bold fashion. If, for example, a wall of natural-looking rock is to be made, why not let it be ten feet high, instead of three feet, and extend it for many yards, instead of a few feet? Steep land and a few rocks will always make one think of the reasonableness of a seemingly small rocky cliff being naturally fashioned at this point. This shows one of the ways in which a new landscape can be actually constructed; the landscape architect then assumes his true function, undertaking always to work on Nature's lines and suggestions, and never in directions uninspired by the existing conditions.

The same principles may be seen applied in the rocks disposed along a stream. If rocks abound, the landscape architect will throw out promontories, picturesque masses and rough bridges, but, if true to his art, he will not undertake to set up cut-stone copings, and long, straight lines, or mathematical curves. It is so easy to say just what should not be done, but the true course is that which seeks to do things right, and forgets to dwell on wrong things. It is the same sensible way of doing right things that leads one to use plain, solid steps of granite or bluestone, and as few of them as possible, instead of searching far and wide for some more rustic or shaly pieces. These steps should be broad and low, and not highly axed and smoothed, and the borders should be, in most cases, improved with low, narrow, inconspicuous copings or curbings, and bordering turf masses, well rounded up.

So, also, should the rocks be selected, being taken from the nearest quarry or stone-heap, without much regard to their shape or size, except that they should not generally be less than two feet the narrow way, with as shallow a depth as possible, so that they will bed down in the ground in what looks to be a firm and settled manner. Although it does not really make much difference whether the upper surface of the rock be weather-worn and moss-grown, these qualities are good to have, and yet good rock-work can be made without them. There is a faculty whereby experience enables one to select a very common-looking stone, and make it do excellent work in the landscape. Indeed, it is, perhaps, easier to select a suitable stone than it is to place it properly after you have selected it. When one feels, as it were, the folds and shape of the land and the way

BRIDGE OF BOULDERS

the rock disposes itself in the neighborhood, it becomes possible to set stones at right angles with the land, to tilt them up, or to bed them lightly or deeply, as occasion requires, and to generally group and mass them, and combine them with grass and vines and shrubs and trees, until a consistent picture is presented wherein the forced note and unnatural quality have absolutely disappeared. This may be done by means of one great rock, or boulder, set at the base of a slope, or by means of hundreds of medium-sized fragments of stone, clothing a hillside, where the presence of a barren, rock-strewn region suggests itself. It is to be noted that rocks among ferns may be made effective in woodland or in some low, moist dell only.

BRIDGE OF BOULDERS, WITH ROCK TREATMENT OF STREAM, IN CENTRAL PARK, NEW YORK

Again, it is a good idea to remember that we can construct rightly and effectively, in the boldest way, bridges and footways of rock, always provided we do it sensibly, in regions where rocks crop out and footways are needed.

A specially happy effect of this kind has been contrived in the upper part of Central Park, New York City, where a lot of great boulders, weighing tons each, have been brought together so as to make a solid arch, under which a stream of water, with heavy rocks strewn on its edge, passes. The region is a deep gorge with high, rocky, wooded slopes, and a carriage-road that passes over the arch gives a sufficiently good reason for its construction. The result is beautiful and natural, although unexpectedly picturesque.

ROUGH STONE WALL AND COPING

If water can be brought directly against a sheer mass of rock five or ten feet high, the effect of rock and water is generally fine. In the same way, artificial constructions of dry walls of rocks, used as dividing barriers between different parts of the property, can be employed in connection with vines and shrubs in an entirely satisfactory way. But cemented stone-work, employed in formal lines, should properly be restricted to buildings or terraces. It is a mistake to underrate the difficulty of doing good landscape work with rocks, and though probably few people realize it, it is a fact, nevertheless, that it is always much better to altogether avoid employing rocks, unless we are willing to spend much time and study on their disposition, and are sure that we can accomplish, in the end, something really good,

TREATMENT OF STEPS WITH ROCKS, CENTRAL PARK, NEW YORK

because it is seldom that the surroundings are such that a reasonably satisfactory landscape *cannot* be devised without the use of rocks. Plants, herbaceous plants, vines, shrubs, trees, and grass we must have, with paths and roads, but the rocks can generally be left out without serious loss, and their absence is better than their presence if their deportment in the landscape is not all that it should be.

There is also a purely conservative side to rock-work that deserves the most careful attention in laying out and constructing a country place of the smallest dimensions (and the smaller it is, the more important), and that is, the preservation, religiously, of rock masses in their natural moss-grown and water-worn state.

Every well-placed, interesting stone in a place has a distinct value in the landscape effect of the region that is difficult to overestimate in actual money. It was the realization of this fact that made a gentleman, known to the writer, say that he would not have sacrificed a certain great, moss-grown boulder, ten feet in diameter, standing on the edge of a spring, for the sum of a thousand dollars, if a thousand dollars would have saved it from the sacrilegious hand and dynamite of the cellar-builder.

RESIDENTIAL PARKS

THE arrangement of lots on a territory that is intended to be divided for residential purposes amounts, in the minds of many, to merely making lots of three-quarters or half an acre, or less, as the demand may require, by running parallel lines, crossing them at right angles; the parallelograms, with a view to tempting purchasers of small means, being sometimes reduced to the normal city lot, size 25 x 100 feet, the roads being worked in between the lots in the same straightforward, simple fashion. This method of laying out villa sites and residential parks has the advantage of economy and simplicity, and the territory certainly divides up into small parcels conveniently; but how about such a simple arrangement from an æsthetic point of view, when the ground is rolling, or even mildly precipitous? Is it not to a proper plan what the row-of-boxes-pierced-with-holes style of building is to true architecture?

There are such things as steep grades that need to be overcome, and that, consequently, force roads into curious and perplexing curves. The location of the houses in each lot is, moreover, a matter that requires skill and special knowledge; in other words, experience and study are necessary for the best results. The drainage problem is rarely a simple one, and the character of the road best suited to the territory is a matter that also requires due consideration and study.

To sum up, if the reader wishes to divide up his property into building lots, on account of a demand for moderate-sized homes that has sprung up in his neighborhood, he needs to think of a great many things that will be required for the proper development of both the inherent and undiscovered beauty and the usefulness of his property for the purpose to which he intends to convert it.

Perhaps as satisfactory a method of discussing this subject, which is evidently a more fruitful one than it would appear to be at first sight, is the consideration of an actual example we have before us, on page 180, where nearly all the problems that would have to be faced on any place of the kind seem to confront us in one form or another. The property in question is a bit of hilly country, in an inland town in the South, and is picturesque and charming to a high degree, being clothed in part by a beautiful variety of forest trees—oaks, chestnuts, etc.—and, at the same time, looking out from its more open portions over a lovely mountain landscape. People come to the region both winter and summer for the enjoyment of the climate and the beautiful natural scenery, and although there are several large, luxurious hotels and country estates, there is a great want felt for houses and grounds of moderate dimensions and expense.

With a view of supplying this demand, the owner of the property we are considering has undertaken to divide its thirty-five acres into lots of an acre or two, and to lay out convenient roads that will reach all parts of the grounds, and that will be specially adapted to securing the best outlooks and vantage points for the scenery. Lots of medium size, from one-half an acre to an acre, had to be secured, and the best places for houses on them suggested. The problem was a knotty one, and one that depended largely for its difficulty on the irregularity and picturesqueness of the contours. The place had fine views, and not much else that fitted it for a residence park. If the owner had fully realized in the beginning all he would have to contend with, it is doubtful whether he would have deliberately faced the difficulties. It would have been so much easier to select a level or rolling piece of ground, where the grades would have been reasonably easy, and the course of the roads and the shape of the lots so much more readily adapted to the ends of design. It would have been altogether so much more satisfactory and sensible. But then undue consideration of beautiful objects will always tend to lead us away from the paths of wisdom, and the residential park-maker did really love those mountain views; and then it must be remembered that gently rolling meadows and level plains did not abound in front of those mountains. Let us look at the conditions that it was necessary to face. Here was a piece of ground (look at the section on page 182, and it will be evident) where the grade from the lowest point to the highest is over twenty per cent., or a rise of one foot for every five feet of longitudinal extent. In other words, one end of a forty-acre plot is 300 feet higher than the other; and to make the problem still more difficult, the contours are strongly plicated laterally, so that a backbone or ridge runs right up through the center, with deeply depressed valleys on either side (see page 181.)

Over at least half of this territory a fine forest of chestnuts and oaks extends itself. It is not very wonderful, therefore, to discover that the mountain views from these slopes are enchanting on a bright, sunshiny afternoon; but, oh, the mountain torrents that tear down through these valleys, and the perched-up sites for houses that seem to offer insoluble problems to those whose unhappy lot it is to devise reasonably accessible houses by means of devious roads and paths. The first thing done was the establishment of a drainage system that was large enough to take care of the greatest rush of water that it would be possible to imagine would at any time occur. The pipe used was in part two feet in diameter, diminishing gradually to one foot, and extending throughout the lowest part of the valleys the length and breadth of the territory.

At the entrance of the park, at the lower end, a simple but picturesque stone lodge, with an arched passage through, has been erected, and here business offices are located. Just within the entrance, a small reservation, some three acres of open space, has been reserved for a small hotel or casino, and lawns with trees. This building is low and picturesque, and is reached by a winding road, for the passage of which it has been necessary to make deep cuts, in order to partially overcome the original inordinately steep grade of twenty per cent., so far as to secure a fourteen per cent. grade, over which it is possible for a carriage to pass with some degree of comfort.

It is evident that on this tract of land the extremes in the way of grades have been reached. The puffing and blowing produced by the mountain climb must be compensated by the mountain views, which, as we strain upward, we may feel triumphantly are far better than those of the flat and uninteresting plain. The problem to be overcome may be difficult, but it is not surprisingly difficult, and is, without doubt, well worth the trouble of an intelligent study that will satisfactorily develop its possibilities for reasonable comfort and beauty. After all has been done, however, that can be done by the most careful study of grades, there still remain spots that cannot, by any possible contrivance, be used for living purposes, unless it be proposed to assume the habits of an eagle on a crag. The only thing that can be done is to at least delight the eye on these inaccessible spots with a thick covering of trees, shrubs, and vines.

A little open meadow and moderately sloping hillside is retained near the entrance, where only is to be noticed any considerable stretch of turf for greensward.

The whole region is a mountain hillside, with trees, shrubs, and vines largely clothing its slopes, and therefore the intention is evident everywhere of supplementing the work of Nature in the same spirit, but with a distinct view of making tasteful and comfortable human homes within its confines. With this object in view, the roads are built solidly, with macadam foundations, and graveled, and all connected by solid stone gutters and road basins with the general drainage system. Occasional hillside flights of steps are introduced, to reach house sites that a carriage may not attempt to approach. Retaining walls along roads also have been found necessary. But the unique, and specially important, adornment of the territory is the plantations that are everywhere carved out on a distinct system. They are really a development of the irregular and natural masses of foliage that already exist on the place, keeping distinctly in view the open-air comfort that is needed for each and all of the houses. It might seem, at first sight, that little planting would be needed in such a thickly-wooded region, but it is astonishing how many plants can be used in what is already a comparatively well-clothed territory.

In the first place, along the drives, wherever there is not ample shade from existing forest, there is a sufficient number of trees to temper the rays of the sun, not necessarily regularly planted, or planted alone, without the association of shrubs, but brought in as a shelter about every forty or fifty feet. They consist, chiefly, of American ashes, tulip trees, American lindens, pin oaks, chestnut oaks, wild cherries, and one or two kinds of maples, and the Oriental plane tree. On the rugged reservations not fitted for residential purposes, shrubs like the lonicera fragrantissima, spiræa opulifolia, forsythia fortunii and suspensa, itea virginica, symphoricarpus glomerata, philadelphus, red-twigged dogwood, ligustrum sinensis, one or two Japan elæagnus, and a Japan barberry. These are not all that have been used, but they are the most important, because they will associate themselves well with the vegetation of the region.

The crowning improvement of these plantations, however, will be found in the vines and creepers that appear everywhere along the roads, over the rocks, and down the steep banks where a goat could hardly climb. The peculiarity of the vine treatment of this small park is that it is used along the roads, because the grass turf will not thrive as well on these steep banks, or harmonize as well with the rugged character of the scenery.

The planting is made with such vines as honeysuckle, running roses, wistaria frutescens, English ivy, etc.,

180 *HOW TO PLAN THE HOME GROUNDS*

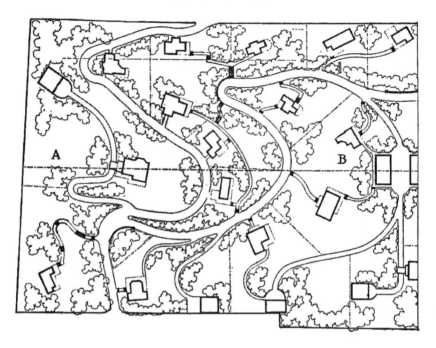

ALBEMARLE PARK, ASHEVILLE, N. C., SHOWING

planted one foot to two feet apart, and extended in an irregular border up and down, and along the base of the steep banks, and along the edge of the road. This treatment is charming in its wildwood effect, and is, in the long run, economical. The vines, to get an early and good effect, should be planted about a foot apart, although even at three feet apart they will make a thick turf, or mat, in two or three years.

In many places English ivy will do well used as turf on the ground, and is certainly very picturesque employed in that way. No vine, however, is better suited to this hillside territory than the Michigan running prairie-rose, rosa setigera. As it has been noted elsewhere, its growth is vigorous, its foliage healthy, and its bloom most profuse.

Even in the lots themselves, a few trees and shrubs, as

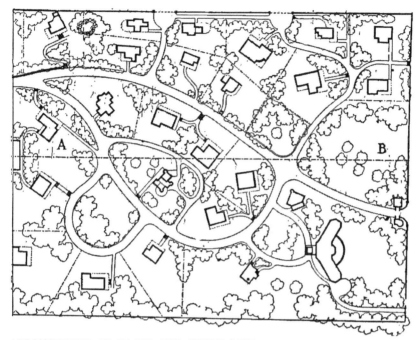

ARRANGEMENT OF ROADS AND HOUSE LOTS

well as vines, are planted in harmony with the general system, so as to establish a certain standard of planting, in the hope that it may be kept up in the future

CROSS SECTION OF ALBEMARLE PARK, ASHEVILLE, N. C.

by all those who buy lots on which to construct homes.

The entire exterior boundary of the place has a stout wire fence, covered with honeysuckles and Virginia creepers, and wherever trees and shrubs are lacking, care

LONGITUDINAL SECTION OF ALBE

is taken to fill in foliage by planting kinds that harmonize with the landscape of the region.

The sites of the houses are carefully selected, so as to get the best views, and several of them have been built and sold by the company controlling the park, so as to set the style of a high and suitable grade of architecture. The treatment of one piece of reserved land deserves special attention, where a bowl, or hollow, has been made by the course chosen for the road. A few small oak trees stand here, but the beauty of the spot has been specially improved by an undergrowth of vines and creeping evergreen plants, and the introduction of a noteworthy collection of the splendid native American azaleas: azalea calendulacea, azalea vaseyii, and azalea viscosa, some of which glow in May and June with the most splendid tints of orange and red. On the park reservations rustic seats and summer-houses are introduced, with all the appurtenances for different games: tennis, golf, croquet, etc.

This entire territory is a particularly difficult place to treat, and yet every spot has always its own special difficulties to meet, whether it be of drainage, sites for the houses, direction of roads, on account of grades, or subdivision of lots.

MARLE PARK, ASHEVILLE, N. C.

It may seem at first sight easy, but there are really a great many points to be adjusted to the æsthetic and practical needs of residents before the arrangements of a residential park, for that is what a building-lot scheme should be, can be considered completely successful.

Its treatment need not be surprisingly original, or fanciful, or picturesque, but there should be such sane consideration of all aspects, practical and æsthetic, of the possibilities of the case, as will secure that sort of perfect relation of all its parts which will give it a dignified and sensible beauty that, if it does not surprise at first, will charm after all, and will *last*.

FENCES, BRIDGES, AND SUMMER-HOUSES

GENERALLY speaking, the buildings on a country place belong to the domain of architecture, having only the associated interest in the landscape that grows out of their juxtaposition. There are, however, a class of structures thoroughly architectural in their character, but having such close sympathy with their environment that it is easy to feel that they belong to the landscape in a more intimate way than, for example, house and stables. These structures exist because they are a distinct development of some special landscape feature with which they harmonize, and which they complete.

Fences, bridges, and summer-houses would never come into being if the boundaries, streams, and foot-paths of the country place did not exist; but the house and stables do not primarily owe their being to any landscape, for man and horse alike must have covering from the weather, though, in place of lawns and groves, there should be stone-paved courtyards.

Perhaps the most difficult problem to solve of this landscape form of architecture is the management of fence or wall. It is a perfectly simple proposition that fences or walls should exist in some shape, wherever boundary lines are established. There is no doubt, on the other hand, that our landscape would be almost invariably improved if we could eliminate the wall or fence altogether, for, as with roads and paths, the landscape would be better without than with them. There is nothing in the presence of generally straight walls cutting across the line of vision that can make them altogether acceptable, but the length to which they are extended may be, by exercise of ingenuity, limited to the shortest possible distance, and the design of the wall constructed may be greatly developed and improved in the direction of agreeable lines and masses of color; and, further, its objectionable character may be greatly suppressed by sinking or screening it, allowing the eye to pass over it, as in the case of the so-called ha-ha fence (see page 186), with its ditch six to ten feet wide, and a dry wall two and a half feet to three feet high on the side of the excavation nearest the pleasure ground which is to be protected from the farm.

Indeed, a stone or brick wall may have a rounded, and even graceful, cap or top line, and a surface with an attractive amount of light and shade, produced by suitably designed recesses or roughness in the material used, and, above all, it may be made charming by English ivy, planted on the north side, or, better still, by Japanese ivy, which should not be mingled with Virginia creepers and the roses setigera and wichuriana, for it is well to keep, in the main, the different kinds apart, and to limit the climbers on such walls to the more strictly decorative sorts just mentioned. The wilder, more unrestrained, and less dignified kinds find fitting positions on low stone walls, lattice-work, and the lawn bank that slopes to the top of a wall and then drops to the side of the walk or road below. It is interesting, moreover, to see how Nature will work on these walls of brick or stone, in the course of time, by painting their surface with minute mosses or lichens, and weather-beaten stains, and as it is seldom that man, with all his vaunted skill and his use of vines, can accomplish altogether satisfactory work in this direction, his best course is, after all, to leave his walls largely to the slow action of the varying seasons. It will surprise many readers to find how great a source of pleasure walls, generally so devoid of interest, become to them, as they

HA-HA FENCE, FOR SEPARATING PLEASURE GROUNDS FROM FARM LANDS

learn to develop, and bring out the possibilities for beauty, of the plainest kinds. When, moreover, they are once satisfied that walls are worth having for their own intrinsic beauty, they will begin to realize that, if a barrier can be made so attractive, the seclusion produced by such features is a good thing to have. Of course, one may object to being shut in by a wall ten feet high, as some of the English parks and even moderate-sized English homes are, but one likes to be more or less by one's self, and not *on gaze*, in order to properly rest or enjoy peaceful recreation.

Surely if we are going to spend much time on our home grounds, there does not seem to be much pleasure

FENCES, BRIDGES, AND SUMMER-HOUSES 187

or advantage in being in *evidence* all the time. Perhaps the reason some people in America have been attempting to do away with the fence and stone wall altogether, and in their place to carry the lawn directly to the sidewalk, is because the stone wall fails to interest them sufficiently, and they desire, instead, to obtain a greater feeling of freedom; but we believe that if they learned how to design the form and coloring of the wall better, and to ornament its surface with suitable vines, we would not hear so much about making lawns without wall or fence. The author undoubtedly knows residence

IRON PIPE AND ANCHOR-POST FENCE

portions of some Western towns where the absence of walls on the sides and front is a most attractive feature, but, as a rule, we are concerned with country places where the requirements are better met by walls and fences.

In favor of the erection of fences there is less to be said, though their use is sometimes imperative; fences are an advantage to the place only in so far as they afford the seclusion and protection from without. They are less defensible than walls, because they can hardly be made altogether satisfactory in line, contour, or surface; therefore, what we can do with fences to render them tolerable is to either make them a solid barrier of close-growing vines, or to construct them of wires which, at a little distance, are entirely invisible. Such fences may be made successfully of locust posts joined by bars of inch gas-piping, or they may be made of wire and small wooden stakes, or iron anchor-posts.

In the eyes of many, the picket fence, in all its forms of both iron and wood, is a contrivance that tends to produce a disagreeable effect, because its upright bars are apt to multiply and confuse the detached glimmers of view we get through the regularly intermittent open spaces. If the picket fence is covered largely with vines, as the horizontal bar form ought to be, there can be little objection to it. In this respect it is easy to recognize the superiority of the stone wall when partially vine-covered, for the variety of contrast between the surface of the stone wall and the leaves of the vines—and there should always be stone wall exposed—will make an effect of changing beauty that a vine-covered fence cannot hope to equal.

It is impossible to commend too highly the use, along walls and fences, of what may be termed hedge-rows, and which are actually shrub-groups. Hedge-row seems to be a good term to employ, because it indicates a certain wild character that greatly increases the attraction of masses of shrubs and trees along a stone wall or fence, and although this wild appearance has been probably secured by setting afresh native shrubs, or kinds allied to them in habit, and planting them with considerable art, it is the chief advantage of the arrangement that the shrubs will seem to have grown there themselves, and not to have been imposed, just at that point, as an artificial barrier.

If it can be so arranged, it is a good idea to secure variations in the seclusion of hedge-rows, or shrub and tree borders, by leaving out trees at considerable intervals, and thus securing more distant and agreeable views which, in any case, are valuable for the change they give to the general scope of the scenery. There is naturally, no ordinary limit to the small size of the place or village lot where this hedge-row, or border, can be used effectively, because three trees and a dozen shrubs of the

SUMMER-HOUSE, CENTRAL PARK, NEW YORK.

right kind, and rightly arranged, will make, in their way, as satisfactory and agreeable a screen for the hard lines of the fence or stone wall, as if they were replaced by a bordering of shrubs and trees a mile long. Naturally, the normal size of the kinds of shrubs and trees used on narrow, long lots, fifty or seventy-five feet by a hundred and fifty feet, should be much smaller, since the scale of everything is smaller.

The problem of designing summer-houses, arbors, and what are termed rustic buildings of different kinds, and setting them at suitable points on the lawn, naturally

190 *HOW TO PLAN THE HOME GROUNDS*

proves attractive to all lovers of picturesque effects. As ordinarily seen, rustic architecture of this kind means the use of white cedar covered with bark, and cut into many short pieces that are worked into the design of a form of lattice-work, that often proves to be ingenious rather than attractive; and, furthermore, this kind of rustic wood displays an unfortunate tendency to drop its

SUMMER-HOUSE

bark in great pieces, owing to decay setting in under the bark and on the upper surface of the wood. Naturally, this decay is liable to extend, in a comparatively short time, throughout the whole structure, so that it will be apparent at once that all bark should be stripped off before using. Instead of white cedar, it is better still to employ red cedar, or the yellow locust, one of the most enduring of woods, for this purpose, entirely stripping it of bark, and setting it in the ground on brick or stone foundations.

Used in this way, in accordance with simple and tasteful designs, rustic architecture finds its proper place in the domain of the home grounds, but the moment the rustic building becomes more pretentious in appearance than is needed to carry the clinging folds of climbing vines—and the fact should by this time have been made evident that the structure's special function is supporting vines and giving people convenient and attractive resting places—then such buildings, overloaded and excessively ornamented, look like intrusions on the lawn.

SUMMER-HOUSE, CENTRAL PARK, NEW YORK.

It scarcely needs to be said, however, that the highest designing ability may be exercised profitably on the simple lines and structure needed to complete the plainest arbor or summer-house for vines. There is a little precaution to be taken in training vines over arbors and summer-houses which, while it seems hardly worth mentioning, in practice helps the development in a short time of an effective growth of leaves, and that is the invariable erection of wires or lattice-work on all buildings before the climbing plants are set out. In this way the vines will gain support at once, and push up with redoubled vigor, which may be still more enhanced by daily training of wandering tendrils in the way they should go. There are not many places where summer-houses and arbors find appropriate location, and one of the most important considerations that should govern the selection of locations for such buildings is seclusion. This is the reason the summer-house, or arbor, seems to fit so well in the nooks and corners of flower-gardens.

There are few private places where bridges are needed, and parks are apt to be too much overloaded with them

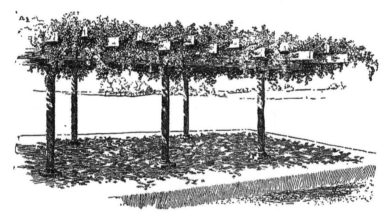

PERGOLA, OR OPEN VINE-COVERED ARBOR

for the general beauty of the locality, although their evident necessity may often force their employment in places where their presence is not, for æsthetic reasons, altogether satisfactory. It is not the intention of the author to make any special suggestions in regard to the actual design of bridges, for that will depend on the situation. In a general way, it may be said that all bridges in ordinary landscape architecture should, particularly when they stand entirely beyond the strict domain of the house, be kept simple, unobtrusive, and subordinated to the primal elements of the home grounds, the dwelling, the lawn, and the plantations. Indeed, it should be always remembered that rustic architecture is a feature of the grounds which should never be introduced except when, as in the case of roads and paths, its practical advantages become evidently paramount.

It is well to bear in mind always, in designing alike

PLAIN RUSTIC BRIDGE IN GENTLEMAN'S COUNTRY PLACE

the arrangement of the smallest village lot and the finest country place or public park, that the presence of a wood, stone, or brick structure will, in the nature of things, produce a certain dissatisfaction when we find that we must accept it in place of more natural objects, for the simple reason that no architect can, in the humble opinion of the author, design a building that is in itself

194 *HOW TO PLAN THE HOME GROUNDS*

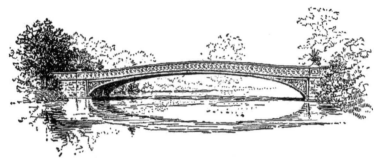

BOW BRIDGE, CENTRAL PARK, NEW YORK.

as attractive as trees and shrubs and flowers and grass. That is why, as in all constructions, the style of the bridge should not be left to the architect alone, but its location, style, and its decoration with trees and shrubs should be decided by the landscape architect.

STONE BRIDGE OVER SMALL STREAM

LIST OF PLANTS FOR GENERAL USE ON HOME GROUNDS

(For description of these plants refer to page references herewith given.)

DECIDUOUS TREES SUITABLE FOR SPRING EFFECT

* Plants distinct and specially suitable for mass planting on the lawn.

Acer campestre, English Field Maple, 96.
Acer rubrum, Red or Scarlet Maple, 95.
*Alnus incana, 107.
Betula alba, White Birch, 3, 65, 103, 124, 245.
Cerasus japonica pendula, Japan Weeping Cherry, 102.
Cerasus sinensis flore plena, 102.
*Cornus florida, White Flowering Dogwood, 109, 110, 244, 245.
Magnolia conspicua, Chinese White Magnolia, 65, 99.
Magnolia stellata, Hall's Japan Magnolia, 99.
Pyrus coronaria odorata, Flowering Apple, 102.
*Pyrus baccata japonica (parkmanii), 65, 102.
Pyrus malus flexilis spectabilis, Chinese Double White-flowering Crab, 102.
*Salix laurifolia, Laurel-leaved Willow, 108, 230.
Salix aurea, 108, 230.
*Salix regalis, Royal Willow.
Salix rosmarinifolia, Rosemary-leaved Willow.

DECIDUOUS TREES SUITABLE FOR SUMMER EFFECT

Acer sieboldii, Japanese Maple, 96.
*Acer platanoides, Norway Maple, 94, 244.
Acer polymorphum, Japan Maple, 3, 65, 96, 244.

Acer atropurpureum, Dark Purple-leaved Japan Maple, 96.
Acer sanguineum, Blood-leaved Japan Maple, 96.
Acer platanoides schwerdleri, 244.
Acer pseudo-platanus, European Sycamore Maple, 95, 244.
Æsculus rubicunda, Red-flowered Horse-chestnut, 65, 102.
*Amelanchier japonica, 113.
*Amelanchier botryapium, 113.
Betula fastigiata, European White Pyramidal Birch.
Betula papyracea, Paper or Canoe Birch.
Carpinus americana, American Hornbeam, 65, 105.
Castanea americana, American Chestnut, 106.
Catalpa bungei, 230.
Catalpa bignonioides, 102, 230.
Celtis occidentalis, American Nettle Tree, 113.
Cerasus serotina, Wild Cherry, 179, 230.
Chionanthus virginica, White Fringe, 114.
Cladrastis tinctoria, Yellow-wood, 101.
*Cratægus crus-galli, Smooth-leaved Thorn, 120, 230.
Cratægus coccinea, 120.
*Cratægus tomentosa, 120
*Fagus ferruginea, American Beech, 65, 104, 226.
*Fagus sylvatica, European Beech, 65, 104, 226.
Fagus heterophylla, Fern-leaved Beech.
Fagus pendula, Weeping Beech, 104.
Fagus purpurea, Purple-leaved Beech, 104.
*Fraxinus americana, American Ash, 103, 179, 226, 244.
Fraxinus juglandifolia, Walnut-leaved Ash.
Fraxinus ornus, European Flowering Ash.
*Gleditschia triacanthus, Honey Locust, 100, 230.
Glyptostrobus sinensis pendula, Chinese Weeping Cypress.
Gymnocladus canadensis, Kentucky Coffee Tree.
Juglans nigra, Black Walnut.
Kœlreuteria paniculata.
Larix kæmpferi.
*Liriodendron tulipifera, Tulip Tree, 98, 179, 244.
Magnolia acuminata, Cucumber Magnolia.
Magnolia macrophylla, Great-leaved Magnolia, 99.
Magnolia watsonii, 99.
Morus alba, White Mulberry.

DECIDUOUS AND EVERGREEN TREES

Phellodendron amurense, Chinese Cork Tree.
*Platanus orientalis, Oriental Plane, 100, 179, 230, 244.
Populus balsamifera, Balsam Poplar, 101.
Populus carolina, Carolina Poplar, 101.
Populus fastigiata, Lombardy Poplar, 101.
Quercus daimio, Japan Oak.
*Quercus palustris, Pin Oak, 105, 179, 226.
*Quercus prinus, Chestnut-leaved Oak, 106, 179.
*Quercus rubra, Red Oak, 106, 226.
Salisburia adiantifolia, 106, 230.
*Salix vitellina, Golden Willow, 108.
Sophora japonica, 106.
*Tilia americana, American Linden, 97, 179, 244.
Tilia europæa, European Linden, 97, 244.
Tilia alba, White-leaved European Linden, 97, 244.
Tilia dasystyla, 244.
*Ulmus americana, American White Elm, 96, 226, 244.
Ulmus campestris, English Elm, 97, 226, 244.

DECIDUOUS TREES SUITABLE FOR AUTUMN EFFECT

Acer tartaricum ginnale, 65, 96.
*Acer rubrum, Red Maple, 95.
*Acer saccharinum, Sugar or Rock Maple, 94.
*Liriodendron tulipifera, Tulip Tree, 98.
*Liquidambar styraciflua, Sweet Gum or Bilsted, 103.
Quercus coccinea, Scarlet Oak.
*Quercus rubra, Red Oak, 106.

EVERGREEN TREES SUITABLE FOR SUMMER AND WINTER EFFECT

*Abies alcocquiana, Alcock's Spruce.
Abies canadensis, Hemlock Spruce, 123.
*Abies douglasii, Douglas' Spruce.
*Abies orientalis, Eastern Spruce, 122, 226.
Juniperus chinensis, Chinese Juniper.
Juniperus japonica, Japan Juniper.
Juniperus prostrata, Prostrate Juniper.
Juniperus squamata, Scaled Juniper.

*Juniperus virginiana, Red Cedar, 122.
*Pinus austriaca, Austrian or Black Pine.
*Pinus mugho, Dwarf Mugho Pine, 122.
*Pinus cembra, Swiss Stone Pine, 122.
*Pinus strobus, White or Weymouth Pine, 122, 226.
*Taxus cuspidata, 123, 226.
Taxus baccata, English Yew, 226.

DECIDUOUS SHRUBS SUITABLE FOR SPRING EFFECT

*Amelanchier japonica, 113.
*Azalea calendulacea, 109, 182.
*Calycanthus floridus.
Chionanthus virginica, White Fringe, 114.
Cornus alba sanguinea, Red-stemmed Dogwood, 110, 179.
*Cornus mascula, Cornelian Cherry, 110.
Cornus paniculata, Panicled Dogwood, 110.
*Corylus avellana, European Hazel, 110.
Cydonia japonica, Scarlet Japan Quince, 109.
Daphne mezereum, Common Mezereon.
Deutzia scabra, 110.
Diervilla hortensis nivea, White-flowered Weigela, 110.
Diervilla rosea, Rose-colored Weigela, 110.
Exochorda grandiflora.
*Forsythia fortunii, Fortune's Forsythia, 111, 179, 244.
Forsythia suspensa, Weeping Forsythia, 111, 179, 244.
Forsythia viridissima, 111, 244.
*Lonicera fragrantissima, Upright Honeysuckle, 115, 179, 230.
Lonicera standishii, 244.
' Lonicera tartarica, Tartarian Honeysuckle, 244.
Lonicera cœrulea or sullivantii, 244.
*Philadelphus grandiflorus, Mock Orange, 116, 179, 230, 244.
Pæonia moutan, Tree peony.
Philadelphus laxus, 116, 230, 244.
Philadelphus tomentosa, 116, 230, 244.
Prunus triloba, Double-flowering Plum.
Ptelea trifoliata.
Pyrus arbutifolia.
*Spiræa thunbergii, 117, 135.
Spiræa van houttii, 117.

Viburnum opulus, 118.
*Viburnum opulus sterilis, Guelder Rose, 118.
Viburnum plicatum, Plicate-leaved snowball, 118.

DECIDUOUS SHRUBS SUITABLE FOR SUMMER EFFECT

Aralia pentaphylla.
*Azalea nudiflora, Pink-flowering American Honeysuckle.
*Baccharis halimifolia, 117.
*Berberis canadensis, American Barberry, 112, 244.
Clethra alnifolia, Sweet Pepper Bush, 109.
*Elæagnus longipes, 111, 179, 230.
Euonymus atropurpureus, Burning Bush, 111.
Hibiscus syriacus, 114, 230.
Hydrangea quercifolia, Oak-leaved Hydrangea, 230.
Hypericum aureum.
Hypericum calycinum.
Hypericum kalmianum.
*Hypericum moserianum.
Itea virginica, 109.
*Ligustrum vulgare, Box-leaved Privet, 74, 115.
Ligustrum ovalifolium, California Privet, 114.
Ligustrum sinense, Chinese Privet, 179.
*Ligustrum ibota, 115, 244.
Myrica cerifera, 117, 230.
Ptelea trifoliata.
*Rhodotypus kerrioides, 3, 116.
Rhus aromatica.
Rosa rugosa, Japanese Rose, 119, 244.
Rosa rubiginosa, 57, 119.
Rosa multiflora, 119.
Rosa carolina, 57, 119.
Rosa laxa, 119.
Rosa, Sweetbrier, 57, 119.
Spiræa ariæfolia.
*Spiræa opulifolia, 117, 179, 244.
Spiræa van houttii, 117.
*Symphoricarpus vulgaris, 3, 117, 179.
Syringa japonica, 110.
Syringa japonica ligustrina pekinensis, 110.

Syringa villosa, 110.
Tamarix africana, 230.
Tamarix chinensis, 230.
*Viburnum dentatum, 119, 244.
Viburnum sieboldii, 118, 244.
Viburnum opulus, 118, 230, 244.
Viburnum plicatum, 118.
Viburnum lantana, 118, 244.
*Viburnum prunifolium, Plum-leaved Viburnum, 118, 230.

DECIDUOUS SHRUBS SUITABLE FOR AUTUMN EFFECT

*Andromeda (oxydendron arborea), 108.
*Berberis thunbergii, 112, 179, 244.
Euonymus europænus, European Euonymus, 111.
Hydrangea paniculata, 230.
Hydrangea paniculata grandiflora, 114, 230.
Rhus cotinus, 113, 230.
Rhus osbeckii, 230.
Rhus typhina.
Spiræa thunbergii, 117, 135, 244.

EVERGREEN SHRUBS SUITABLE FOR SUMMER AND WINTER EFFECT

Andromeda catesbii, 109, 128.
Azalea amœna, 126.
*Buxus sempervirens, Tree Box, 128.
Cratægus pyracantha, 126.
Ilex opaca, American Holly, 74, 125.
Ilex crenata, Japanese Holly, 74, 125.
*Kalmia latifolia, Mountain Laurel, 127.
Mahonia aquifolium, Holly-leaved Mahonia, 126.
*Rhododendron catawbiense, Catawba Rosebay, 127.

DECIDUOUS CLIMBERS SUITABLE FOR SUMMER EFFECT

Actinidia polygama, 143.
Akebia quinata, 226.
*Ampelopsis quinquefolia, Virginia Creeper, 142, 185, 226, 245.
Ampelopsis tricuspidata (veitchii), Japan Ivy, 141, 185, 226, 227.
Aristolochia sipho, Dutchman's Pipe, 143.

HERBACEOUS PLANTS AND ANNUALS 201

Celastrus scandens, Climbing Celastrus.
Clematis henryii, 141.
Clematis jackmanii, 141, 226.
Clematis hybrida sieboldii.
*Clematis paniculata, 141, 245.
Clematis virginiana, American White Clematis, 141.
Dolichos japonica, 143.
Hedera helix, English Ivy, 143, 185, 227.
Lonicera brachypoda aureo reticulata, Japan Golden-leaved Honeysuckle.
Lonicera sinensis, Chinese Honeysuckle, 142.
*Lonicera halleana, 226, 245.
Periploca Græca, 141.
*Rosa setigera, Michigan Climbing Rose, 3, 143, 226.
*Rosa wichuriana, Memorial Rose, 3, 144.
Tecoma radicans, American Trumpet Creeper, 142, 226.

DECIDUOUS CLIMBERS SUITABLE FOR SPRING EFFECT

*Wistaria sinensis, Chinese Wistaria, 142, 226.
Wistaria frutescens, Clustered Flowered Wistaria, 142, 226.

HERBACEOUS PLANTS AND ANNUALS SUITABLE FOR SPRING EFFECT

Aquilegia canadensis.
Aquilegia chrysantha.
Crocus verna, 134.
Hepatica triloba.
Iris pumila.
Iris pseudacorus, 132.
Lily of the Valley, 134.
*Narcissus princeps and major and incomparabilis, 133.
*Narcissus polyanthus, 133.
Narcissus, Poet's Narcissus, 134.
Pansies, 57, 58.
Peonies, single and double, 135, 136.
*Phlox subulata rosea, 133.
Sanguinaria canadensis.
Trillium nivale.

Trillium grandiflorum.
Trillium erectum.
Tulip, Hardy.
*Vinca minor.
Viola pedata.

HERBACEOUS PLANTS SUITABLE FOR SUMMER EFFECT

Acorti calamus.
Antirrhinum (Snap Dragon).
Anemone japonica, 134.
Arethusa bulbosa.
Asclepias tuberosa, Milk-weed, 135, 231.
*Aster novæa-angliæa, 135.
Aster, Annuals, 57.
Boltonia.
Begonia vernon
Callirrhoe involucrata.
Calleopsis, Tom Thumb.
Calopogon pulchellus.
Campanula medium, Canterbury Bell, 135.
Campanula rotunifolia, Hare Bell.
Cassia marilandica, American Senna.
Chrysanthemum filifera, 135.
*Coreopsis lanceolata, 133.
Dahlias, 132, 135.
Delphinium (larkspur) formosum, 57, 70, 130.
Dianthus barbatus, Sweet-william, 133.
Dianthus plumarius (Garden Pink), 57, 133.
Dicentra cucularia, 133.
Dicentra exima, 133.
Dicentra formosa, 133.
Eulalia japonica.
Euphorbia corollata.
Ferns, Hardy, 170.
Gladioli.
Gaillardia, 133.
Gentiana andrewsii.
Geranium maculatum.
*Helenium autumnale, 135, 136, 231.

HERBACEOUS PLANTS AND ANNUALS

Helianthus maximilianii, Sunflower, 57, 70, 135, 136.
Helianthus veitchii.
Hemerocallis flava, Day Lily.
Hibiscus crimson eye, Rose-mallow, 135, 231.
*Hibiscus moscheutos rosea, 135, 136, 231.
Hollyhocks, Alleghany kinds, 57, 135, 170.
*Iris germanica, German iris, 132.
*Iris pseudacorus, 132, 140.
*Iris kæmpferi, Japan iris, 132.
*Lilium auratum, 132.
Lilium longifolium, 132.
Lilium canadense, 132.
Lilium album, 132.
Lilium superbum, 132.
Lilium tigrinum, 132.
Marigold, dwarf, 57.
Mignonette, 57.
Myosotis palustris, 57.
Monarda didyma.
Nasturtium, 57.
Nelumbium speciosum, Lotus, 139.
Nelumbium luteum, 139.
Nymphæa odorata, American Water-lily, 139.
Peas, sweet.
*Phlox, garden, 37, 132.
Phlox drummondii, annual, 57.
*Pontederia cordata, 140.
Poppy, single, annual, large flowered, 57, 136.
Poppy, Iceland, 57.
*Pyrethrum uliginosum, 135, 136, 231.
Rudbeckia grandiflora, 57, 70, 135.
*Sarracenia purpurea, Pitcher Plant, 140.
Solidago cæsia, Golden Rod, 135, 136.
Stylophorum diphyllum, Celandine Poppy.
Tradescantia virginica.
Torrenia fournieri.
Ten Weeks Stock.
Viola canadensis.
Viola cornuta.
*Zinnias.

CONTRACTS AND SPECIFICATIONS

AGREEMENT made this fifth day of November, 1898, between   (hereinafter called the owner), party of the first part, and   of the city of New York (hereinafter called the contractor), party of the second part, WITNESSETH:

That the owner and the contractor, for and in consideration of the sum of one dollar each to the other in hand paid, and other valuable considerations, the receipt whereof is hereby acknowledged, have agreed as follows:

FIRST: The contractor agrees to do and perform all the work and furnish and supply all the materials required to finish and complete the roadways, paths, turf, plantations, mold, top-soil, etc., mentioned and referred to in and called for by the drawings and specifications made by   , landscape artist, and signed by the owner and contractor, relating to the grounds to be laid out on lots belonging to said owner, and known as   in the city of New York, which drawings and specifications are hereto annexed and made a part of this contract, in consideration of the payment to the said contractor by the said owner of the sum of   dollars, at the dates and in the manner hereinafter provided. And the contractor further agrees to perform the work aforesaid in a good and workmanlike and substantial manner, under the direction and to the satisfaction and approval of the said   , landscape architect, his successor or successors.

SECOND: The contractor further agrees that all materials furnished and provided by him shall be of the kind and quality described in said specifications, and further agrees that all materials required to be furnished and supplied which are not described in said specifications shall be of the best quality and shall be approved by said   , landscape architect, his successor or successors.

THIRD : The contractor agrees to begin work under this contract on the first day of January, 1898, and to fully perform, complete, and finish all work, and furnish and supply all materials embraced in this contract and said drawings and specifications, in strict accordance with the terms and requirements of this contract and said drawings and specifications, by the first day of June, 1898, and that in the event of his failure so to do, he will pay to the party of the first part, as liquidated and stipulated damages, and not by the way of penalty, the sum of fifty dollars ($50) for every day that the work called for by this contract shall remain uncompleted and unfinished subsequent to June 1, 1898, up to and including the fifteenth day of July, 1898, and will pay to the owner the sum of seventy-five ($75) a day as liquidated and stipulated damages, and not by way of penalty, for every day that the work called for by this contract shall remain uncompleted and unfinished to said fifteenth day of July, 1898 ; and the said contractor further agrees that the said owner may deduct and retain said liquidated and stipulated damages out of any moneys due him under the terms of this contract at the date of said damages, or any part thereof, shall accrue, or out of any moneys that may thereafter become due to said contractor under the terms of this contract.

FOURTH : The contractor agrees that he will, at his own expense, provide and furnish any and all materials (including water), labor, tools, implements, and cartage, of every description, necessary to the due performance of said work and the full and complete performance of this contract.

FIFTH : The contractor further agrees that he proceed with the said work, and every part and detail thereof, in a prompt and diligent manner, and at such reasonable times as may be necessary and proper in order to complete and finish the same, and every part and appurtenance thereof, in a durable and substantial manner, on the said first day of June, 1898, and without the performance of any part of the said work in unsuitable weather.

SIXTH : The contractor further agrees that he will not at any time suffer or permit any lien, attachment, or other incumbrance, under any law of this State, or otherwise, to be put on the premises upon which the aforesaid work is to be done, and for which the aforesaid materials are to be furnished under this contract for such work or materials, or by reason of any claim or demand against him, the said contractor; and that should any lien, attachment, or incumbrance be placed or filed upon said materials, the said contractor shall not, until such lien, attachment, or other incumbrance shall be removed, satisfied, and discharged, be entitled to claim, demand, or receive any payment whatever under or by virtue of this contract.

SEVENTH : The contractor further agrees that the plans, drawings, and specifications hereinbefore mentioned are intended to coöperate so that any matter or thing contained or shown by one and not by the other shall be of the same force and effect as if contained in and shown by both; and that he will perform any work and furnish all materials shown by either without extra charge, claim, or demand whatsoever.

EIGHTH : The contractor further agrees that the owner may, at any time during the progress of the work, alter, change, deviate from, and add to said drawings and specifications, and that any such alteration, change, deviation, or addition shall in no way affect the validity of this contract. Provided, that if such alterations, changes, deviations, or additions shall decrease the aggregate cost of said work and materials, then the amount of such decrease in cost shall be deducted from the said sum of $, and the said owner shall only be liable to pay to the contractor the balance remaining after making said deductions as aforesaid; and provided further, that if such alterations, changes, deviations, or additions shall increase the aggregate cost of said work and materials beyond said balance remaining after making the deductions aforesaid, then the said owner shall pay to the said contractor the amount of such excess in cost, with ten per cent. on the said excess in addition to said balance.

NINTH : And the contractor further agrees that he shall not be entitled to claim, demand, or receive any pay for extra work, unless the necessity for such extra work shall be certified to by the said   , or his successor or successors, in writing; and the price to be paid for such extra work shall have been fixed and determined before the same shall have been performed, by a written memorandum ordering the extra work to be done, and stating the price to be paid therefor, signed by the owner.

TENTH : The contractor further agrees that if he shall at any time neglect or refuse to supply a sufficient number of workmen of requisite skill, or to furnish materials of the kind and quality called for by this contract and said specifications, or shall fail in any respect to prosecute the said work with promptness and diligence, or shall be in default in the performance of any covenant in this contract contained on his part to be kept and performed, for the period of three days after notice in writing, signed by the owner, of such default shall have been served upon him, either personally or by leaving the same at his residence or place of business, then, and in such event, the said owner shall have the right and power to employ other persons to perform the work and furnish the materials required by this contract, and to complete the same in every respect, and the cost and expense thereof at the reasonable market rates in excess of the unpaid balance of the contract price shall be a charge against him, the said contractor, and he will pay the same to the said owner ; and he, the said contractor, shall have no claim or demand against the owner for said unpaid balance or by reason of the non-payment thereof ; and the said owner, and all persons employed by him to complete the said contract, shall have the use of all fixed tackle of any kind belonging to, or used by, the said contractor prior to said default on his part, free of charge, and until the said contract has been fully performed and completed.

ELEVENTH : The contractor further agrees that the said   or his successor or successors, may condemn any materials furnished, and reject any work performed under this contract, and require the same to be taken up and removed from the premises by, and at the expense of, the said contractor ; and that said   and his successor or successors, may also direct the time of doing the several portions of work called for by the contract.

TWELFTH : The contractor further agrees that no certificate given or payment made under this contract shall operate as, or be held to be, an admission on the part of the owner that this contract, or any part thereof, has been complied with, or that any detail of the work has been properly performed, or that the materials furnished are of the quality called for by the specifications, in case the fact shall be otherwise ; nor shall any such certificate or payment stop or preclude the said owner from claiming damages against the said contractor, should the work and materials hereby required not be performed and furnished in every particular in a substantial and workmanlike manner, and in strict fulfilment and compliance with the requirements of this contract and said drawings and specifications.

THIRTEENTH : The contractor further agrees that he will bear and be liable for all loss or damage that may happen to the said materials by fire, storms, or otherwise, prior to the time they have been actually used and entered into the construction of said grounds, and that he will repair all damage and injury to said work and materials occasioned other than by fire, storms, or otherwise, during the performance of this contract and prior to the completion and acceptance of the same, and without extra charge.

FOURTEENTH : The contractor further agrees to indemnify and save harmless the said owner from all and every claim of damage or injury to person or property occasioned by his negligence, carelessness, or want of skill, or that of his servants or employees, or that of his sub-contractors or their employees, while engaged in the performance of the said work, or otherwise ; and further agrees to indemnify and save harmless the said owner from every claim and demand for the violation by him, his servants, or sub-contractors and their servants, of any statute or municipal ordinance regulating or relating to the work called for by this contract.

FIFTEENTH : The contractor further agrees that in case the said   shall die, resign, be removed, or refuse to act, then the said owner may appoint a successor or successors, and such successor or successors shall have like power and perform like duties as are conferred and imposed upon by the said   by this contract.

SIXTEENTH : The owner agrees to pay to the contractor for performing said work and furnishing and supplying said materials as aforesaid the sum of   dollars in instalments, as the performance of this contract progresses, and as follows : On the twenty-fifth day of each and every month the contractor shall furnish to the landscape architect, or his successor or successors, a statement of the work done and the materials furnished during the thirty days next preceding, and thereupon the landscape architect, or his successor or successors, shall verify said statement and furnish to the contractor a certificate in writing, signed by him, of the value of the work done and materials furnished and supplied and actually used and applied in the construction of said building during said preceding thirty days ; and upon the presentation of the said landscape architect's certificate the said owner will pay to the contractor 85 per cent. of the amount so certified by the landscape architect, and will make such payment within ten days after the presentation of such certificate as aforesaid. When the last instalment shall be certified, the aggregate of the 15 per cent.

deducted from the prior instalments, and the amount of the said instalment, with such additions, shall be paid to the said contractor by the said owner when, and not before, the said contractor has complied with the conditions in the next succeeding paragraph of this contract.

SEVENTEENTH : The contractor agrees that he shall not be entitled to demand, receive, sue for, or collect the amount of said last instalment, or any part thereof, until he has presented to the said owner the certificate in writing, signed by the said   , his successor or successors, to the effect that this contract has been fully completed and performed, and also the certificate of the county clerk of the county of   , that no mechanics' or other liens are of record upon said construction, for work done or materials furnished by any person or persons for, or on behalf of, said contractor or any sub-contractor, or his or their employees, and also only upon evidence being furnished by the contractor satisfactory to the owner, that no claim or demand exists in favor of any person or persons for work done or materials furnished or supplied in the performance of this contract.

EIGHTEENTH : It is further mutually agreed between the contractor and owner that, should any dispute or question arise respecting the true construction or meaning of the drawings or specifications, the same shall be decided by   , his successor or successors, and his or their decision shall be binding and conclusive. But should any dispute arise respecting the cost of any change, alteration, deviation, or addition with respect to the work to be done and the materials to be furnished under this contract, the same shall be submitted to two arbitrators, one to be chosen by the contractor and the other by the owner, whose decision, if they agree, shall be final and conclusive, and who, in case they cannot agree, shall have the power to choose a third arbitrator, and the decision of the three arbitrators, or a majority of them, shall be binding and conclusive upon the said owner and the said contractor.

NINETEENTH : It is further mutually agreed between the owner and contractor that the owner shall not in any manner be answerable for any loss or damage that shall or may happen to the work done, or materials furnished under this contract during the performance thereof, or for any loss or damage that may at any time happen to the materials, tools, and appliances used and employed in the performance of the work called for by this contract.

TWENTIETH : It is further mutually understood and agreed between the owner and the contractor that the contractor shall have the right to make sub-contracts, but only with such person or persons, corporation or corporations, as shall have been first approved in *writing*, signed by the owner, and that the contractor shall not assign this contract or any interest therein without the consent of the owner in writing shall have been first had and obtained, it being understood and agreed that this contract is for the personal service and skill of the said contractor, and that if such contractor shall make any such assignment without such consent, then, at the option of the owner, this contract shall cease, determine, and be null and void.

TWENTY-FIRST : This contract shall bind and inure to the benefit of the owner, his successor or successors, and the contractor, his heirs, executors, and administrators.

IN WITNESS WHEREOF the contractor and owner have caused these presents to be duly executed the day and year first above written.

, Owner.
, Contractor.

CONTRACT WITH LANDSCAPE ARCHITECT

This agreement between   , landscape architect, and   , owner, entered into this   day of

WITNESSETH, That the said   is to draw plans and specifications of grounds to be laid out for the said at   . The drawings are to be complete and to include to scale   inches to foot, together with all other necessary and proper papers and drawings, four copies of each, before 189 , and before taking estimates, in order that each one estimating may know what will be required. The said plans and drawings are to be the property of the said   , owner.

Said landscape architect is further to give supervision to the work throughout its progress, to visit the premises at least once every week and carefully inspect every portion of the grounds. He is to carefully inspect and test all material and finished work, and to protect the interests of the owner in every way.

In consideration for which service the said   is to pay the said   , on the completion of the drawings, plans, specifications, etc., the sum of   dollars, and the further sum of   per cent. on the money due and payable to various contractors under their contracts, said percentage to be payable at the time payments under said contracts are by their terms due and payable.

Witness our hand and seals this   day of   , in the city of
.
(Signed)   (L. S.).
        (L. S.).

SPECIFICATION AND DESCRIPTION OF MATERIALS TO BE FURNISHED AND WORK TO BE DONE IN LAYING OUT AND CONSTRUCTING THE ROADS, PATHS, LAWNS, PLANTING, ETC., FOR THE GROUNDS PERTAINING TO THE RESIDENCE OF MR.   AT

All work to be done in conformity with the plans, details, and other drawings and with these specifications, prepared by   ,
duly employed by Mr.   as inspector to take charge of the work.

DESCRIPTIONS OF DRAWINGS

The drawing referred to in these specifications will be as follows:
(1) General contour plan of property.
(2) General plan showing location of house, roads, paths, steps, terraces, planting, etc., giving the various levels of same.
(3) Drainage map.
(4) Detail drawing of various parts.
(5) Cross sections through various parts, as indicated on plan.
(6) Planting map.

The work must be done in accordance with the above plans and details, explanatory thereof, and such directions and additional detail drawings as may be given during the progress of the work, of
        and inspector duly appointed by him.

In no case shall the contractor measure any scale drawing; in every case where figures are not already given on the drawing, he is to obtain the same from the inspector.

STAKING OUT

The contractor must employ a competent surveyor to locate and properly stake out the entire work as shown on drawings. All grades and levels of finished earth, paths, roads, etc., must be properly marked and otherwise indicated by stakes. The contractor will be held responsible and must make good all damage caused by improper grades. The contractor will notify the inspector when this staking out has been done and will make such changes in same as may be required or suggested by him.

DRIVEWAY

PREPARATION OF ROAD-BED

The subsoil or other matter (be it earth, boulders, tree-stumps, etc.) shall be excavated and removed to such depth as that, when the surface is thoroughly compacted by ramming and rolling, it shall be left   inches below the finished grade.

Should there be any spongy material in the bed thus prepared, all such material shall be removed, and the space filled with clean gravel or sand, and carefully rammed, so as to make all filling compact and solid.

Filling, if required, shall be composed of good, wholesome earth, taken from the adjoining banks, or from where directed, and placed upon the road-bed in layers, of not more than six (6) inches in depth, and thoroughly rolled and rammed. The embankment shall be maintained at its designated grade until finally accepted and no allowance will be made to the contractor for shrinkage or settlement.

Wherever it shall be deemed necessary the exterior of the filling shall consist of a stone wall as per drawing, along the sides of the road, as shall be directed.

FOUNDATION OF ROAD

After the road-bed is properly prepared, sound, durable quarry-stones, about 6 inches in depth, 3 to 6 inches in width, and 8 to 12 inches in length, shall be laid by hand in form of a close, firm pavement, and the various sizes properly distributed. They shall be set on their broadest edges, without underpinning. The interstices are then to be filled with pieces of stone and set with hammers in such a manner that the foundation shall have an average depth of 6 inches.

MACADAMIZING

Upon this foundation a 3-inch thick layer of broken stone, $1\frac{1}{2}$ to 2 inches in diameter, is to be spread and thoroughly rolled, free from clay or earth.

FINISHING

Screenings of broken stone about 1 inch in depth are then to be applied and well saturated with water, and thoroughly and repeatedly rolled, while wet, until a wave is formed in front of the roller.

Should the owner or owners desire to use gravel in place of broken stone and screenings, the contractor is required to submit samples of same to be used in place thereof.

The widths of the road or roads to be as follows :

The gutters to be built either of cobblestones, or pressed composition blocks, or of sod, or as otherwise directed, as shown in section. If cobblestones or composition blocks are used, the same to be laid on a stone foundation (the extension of the road width), as prescribed before, and bedded in a sand cushion.

Provisions for surface drainage to be made as shall be directed, while the work is progressing, either by building stone culverts or laying vitrified earthenware pipes, and setting catch-basins.

Contractors are required to bid on each item separately.

The quantities herein stated are approximate only.

The contractors will be paid for quantities of work actually done.

Engineers' estimates of the work to be done, and by which the bids will be tested, are as follows :

1. ...... square yards of broken stone foundation, as specified before, with 3-inch broken stones and 1-inch screenings on top, per square yard, the sum of...................................$......
2. ...... cubic yards of earth excavation :
...... per cubic yard, sum of..$......
3. ...... cubic yards of extra earth, filling in excess of excavation :
...... per cubic yard, the sum of...................................$......
4. ...... lineal feet of   inch pipe, including laying :
...... per lineal foot, the sum of..................................$......
5. ...... lineal feet of   inch drain pipe, including laying :
...... per lineal foot, the sum of..................................$......
6. ...... lineal feet of   inch drain pipe, including laying :
...... per lineal foot, the sum of..................................$......
7. ...... lineal feet of   inch stone culvert :
...... per lineal foot, the sum of..................................$......

8. ...... cubic yards of rustic retaining walls :
   ...... per cubic yard, the sum of...................................$......
9. ...... cubic yards of rock excavation :
   ...... per cubic yard, the sum of...................................$......
10. ...... catch-basins :
    For 1, the sum of ....$......
    About $20 (as per drawing).
    Payments will be made as follows :
      Twenty per cent. of contract price, when road-bed is prepared to receive road material, and drainage of road provided for.
      Sixty per cent. of contract price, when road is finished.
      Twenty per cent. of contract price, 3 months after road is finished and accepted.
11. The entire work to be done during   working days.
12. The work to commence within one week after the signing of contract.

The parties of the first part shall have the right to increase or diminish the quantities of the several items called for.

SPECIFICATIONS FOR PATHS

PREPARATION OF PATH-BED

The path-beds shall be prepared in the manner specified for road-beds, and when finished shall be left   inches below the finished grade.

FOUNDATION OF PATH

After path-bed is properly prepared, a -inch thick layer of rubble, or broken stone,   inches in diameter, is to be spread and thoroughly rolled, free from clay or earth. Upon this foundation a -inch thick layer of broken stone, or gravel,   inch in diameter, is to be spread and thoroughly rolled, free from clay or earth.

FINISHING

Screenings $\frac{3}{4}$ inch in depth of broken stone, or as otherwise specified, are then to be applied, well saturated with water, thoroughly and repeatedly rolled while wet until a wave is formed in front of the roller.

EXCAVATION, FILLING, AND SHAPING

DRAINAGE

Provisions for surface drainage to be made as shall be directed, while the work is progressing, either by building stone culverts or laying vitrified earthenware pipes, and setting catch-basins.

SPECIFICATION FOR DRAINAGE, ETC.

Provisions for drainage and laying water-pipes, etc. (apart from the drainage specified for roads and paths), to be made according to the lines, depths, etc., shown on drainage map, or as shall be directed while the work is progressing.

SPECIFICATION FOR PREPARING GROUND FOR SODDING OR SOWING AND PLANTING

GRUBBING AND CLEARING

From such portions of the grounds as may be directed all trees, saplings, bushes, stumps, and roots (except such as inspector directs to be saved) shall be cut and thoroughly grubbed up, and together with logs, brush, and wood of every description be removed from the grounds and disposed of by the said contractor. No payment will be allowed for this work of clearing and grubbing, it being considered as included in the price to be paid for excavation.

EXCAVATION, FILLING, AND SHAPING

At such portions of the grounds, and for such widths and depths as may be directed, the soil shall be excavated, and either spread over the ground at such places, and in such depth and to such lines and slopes as may be directed, or placed in piles of such dimensions and at such localities as may be directed, and to give not exceeding   feet haul for said material. These piles to be of regular shape with well-made faces. Said soil to be and remain the property of Mr.   .
The price for excavation of earth per cubic yard to be in full for the excavation, removal, spreading or piling of said soil. If so directed, during the progress of the work, such soil so piled shall be again removed and spread over the slopes or other portions of the work at

such places and in such depth as may be directed, but to give not exceeding   feet haul. For this second movement of soil the price for excavation of earth will be paid, and will be in full for the re-excavation, removal, placing, and spreading said soil.

Where the present surface is above the required sub-grades, the material, be it solid or loose rock or earth, is to be excavated to conform to the required grades, sub-grades, lines, and slopes, and other prescribed lines designed on the ground by the inspector, and the material shall be removed and deposited as filling at such points as shall be directed by the inspector, or otherwise disposed of as hereafter specified.

The materials excavated shall be removed and deposited, as shall in each case be directed, at the points at which it may be required for embankment or filling, or other purposes, and in accordance with the provisions of these specifications. Such materials excavated, unfit, or not used for filling, shall be removed from the work and deposited on the grounds or elsewhere when required, and in the manner directed by the inspector.

The price for grubbing, excavation, filling, and shaping of grounds outside of roadways, paths, etc., must be given in a lump sum.

ROCK EXCAVATION

Where rock occurs it shall be excavated to conform to the required grade, slopes, or other prescribed lines.

The excavation of solid rock, and of boulders or detached rock measuring one cubic yard or more each, will be classed as excavation of rock. No soft or disintegrated rock that could be properly removed with a pick will be allowed for as rock. The excavated rock shall be removed and deposited, as shall in each case be directed, at the points at which it shall be required for its use, as herein specified.

The material excavated shall be removed and deposited, as shall in each case be directed, at the points at which it may be required for embankments or other purposes, and in accordance with the provisions of these specifications, and any surplus not so used shall be removed from the work and disposed of by the contractor as herein above specified.

In all cases of rock blasting the blast shall be carefully covered with heavy timbers, chained together, according to the ordinance of

MOLD OR TOP SOIL

the   relative to rock blasting, and every precaution taken to insure the safety of all persons. The contractor will be required to carefully observe and conform to all the ordinances and regulations of the State of   now in force, or that may be in force during the progress of the work, in relation to the storage and handling of explosives in the streets and avenues of the city or town of

The price for excavation of rock will include the excavation, removal, and final disposition of the material in accordance with these specifications.

FILLING OR EMBANKMENT

The filling or embankment is to be made in accordance with the lines, slopes, grades, and dimensions shown upon the plans, and as shall, from time to time, during the progress of the work, be directed by the inspector.

No rock filling will be allowed within one foot of the finished grades of the roadway and walks, nor in such parts of the filling or embankment as will interfere with trenches or pits to be afterward excavated for basins or pipes.

No rock filling will be allowed in the embankment or filling of the grounds, or of the slopes exterior to lines of the roadway and walks within   feet of the finished surface of the grounds.

Where the filling is required to be two feet and over in depth, if directed by the inspector, it shall be deposited in regular horizontal layers of not more than two feet in thickness, and shall be carted over and rammed.

The materials for the embankment are to be obtained from the excavation of the ground and walks above the required lines, grades, and sub-grades. Any deficiency of the earth for the embankment shall be furnished by the contractor from sources exterior to the grounds of the park at his own cost and expense, compensation for the same to be included in the price to be paid for rock excavation.

All earth filling furnished shall be of good, wholesome earth, free from garbage, vegetable, or other unsuitable matter.

MOLD OR TOP SOIL

After the area of the ground, outside the lines of the roadway, walks, structures, etc., has been properly regulated and graded to the satisfaction of the inspector, the same is to be covered with a layer of mold or top soil of such depth as will bring the top surface of the same up to the finished lines as shown upon the plan, the depth of which will be from   to   inches, with a sufficient allowance for settlement, and shall be evenly spread and leveled to such surface as the said inspector shall direct.

The mold shall be of first quality garden mold or fertile loam, free from stones, roots, and all other extraneous materials, and the quality of the same be to the entire satisfaction of the inspector.

SPECIFICATION FOR PLANTING

The holes for trees and shrubs to be dug sufficiently wide to give   foot of space between the roots spread out and the side of the holes, and sufficiently deep to permit the tree or shrub to stand at the depth in the ground that it stood in the place from which it was last taken. The bottom of the hole must be thoroughly loosened with the spade before it receives the tree or shrub.

SPECIFICATION FOR SEED SOWING

The grass seeds must be the cleanest that the market will afford, and consist principally of Kentucky Blue Grass and Red Top. They must be sown at the rate of six bushels to the acre. The sowing must be done in calm weather, so as to spread it evenly, and when sown it must be carefully raked with a steel rake, and rolled with a heavy iron roller drawn by two horses.

PART II

PARKS AND PARKWAYS

IN considering the general principles that should underlie the arrangement of home grounds, we are not surprised to find that we are covering at the same time the whole theory of the design of parks and parkways, for when we reduce to first principles our several schemes of arranging home grounds and parks, we shall find that we are only, in the one case, contriving pleasure and comfort and good sanitary conditions for the few, and in the other, provided we have substantially the same general conditions of soil and topography, for the many. Instead of one lawn on the village lot, we have a series of lawns in the public park, but the principles of arrangement are actually the same in both places. There are, for instance, in both places open lawns, bordered and framed by plantations of shrubs, trees and herbaceous plants, with as few buildings as will furnish the necessary solace and pleasure for the occupants alike of home grounds and parks, and with these should always be combined a reasonable amount of seclusion. The same adjustment of the lawns and plantations of the home grounds will also make the largest park, multiplied though it be in size a hundred times.

Of roads and paths we have already said that they were necessary, but not in themselves interesting, being simply incidents of connection between different parts of both home grounds and public parks, and, indeed, we

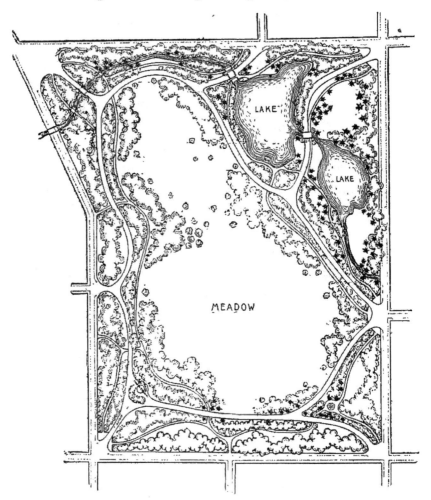

RURAL PARK OF MODERATE DIMENSIONS FOR CITY OR TOWN

must allow that they have no reason for being, except as a plain, practical means to an end. They should approach and pass all the characteristic features of the landscape, in ways that will present the most attractive and beautiful views possible, but they themselves should have massed at their junctions, and various other points, trees and shrubs, to modify the objectionable influence of their uninteresting surface. You may have woodlands, rocky rambles, and straightway lines of paths and roads, but the ultimate analysis of all grouping or arrangements of this kind for all landscape purposes, is the lawn and framing trees and shrubs, the necessary buildings, paths and roads being subordinated and kept as much as possible out of evidence. This helps reduce the problem of park treatment to very simple terms, which will be found to be sensible and sound when they have been duly considered.

In contemplating the different objects that naturally associate themselves in a public park, we find that chief among them are included peace, rest, the means of seclusion, also the suggestion of the best kind of landscape and of country sights and sounds, trees, grass, birds, flowers, open meadows for games for men and boys and little children, and buildings for music, eating, drinking, and all sorts of social intercourse. All of these features are, within certain limits, equally desirable on home grounds of small as well as large dimensions. A park, therefore, is evidently nothing more in its essential character than a great country place where hundreds and thousands may cheerfully resort for the joy of open-air life, for games, and, in secluded places, for the rest and peace of something like sylvan solitude.

These conditions should never be forgotten by those who are seeking to secure parks, small or large, for town or city. It should be, above all things, the undivided aim of every one who may undertake the duty of constructing parks for the public, in the smallest country towns as well as in large cities, to always strive to suggest the country and a country landscape, and to give opportunity for out-door sports.

The author does not wish to be understood as intending, in any way, to underrate the skill necessary to prop-

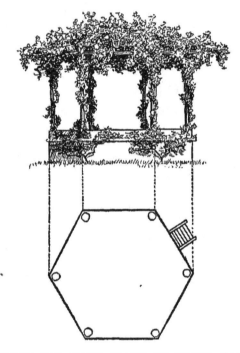

DESIGN FOR BAND STAND IN PUBLIC PARK

erly work out the scheme of a park, nor does he, as will be seen when we consider what it means to perform the delicate operation of modeling a lawn on lines suggested by the original contours, and to plant natural-looking groups on the outskirts and framework of the place, and about the junction of the roads and paths. If we are able to always keep close to these suggestions of country which should be intimately associated with parks, we will find that we will, in that way, shut out all bizarre masses of imperfectly related, inharmonious beds of plants, all detached and unmasked flower-gardens, and bar out, successfully, all advertising schemes, side-shows, race-courses, military displays, and betting resorts, to the end that all people may absorb, undisturbed, the peace and rest and healthful enjoyment that these park suggestions of country life afford.

Amid such scenes child-life expands like a flower, and its innocent games and happiness go on unrestrained by the fear and unrest and constraint that must necessarily exist in all town and city streets, which make the usual play-grounds for many children. Every device in the way of summer-houses, arbors, and seats, and, above all, of police guardianship, should be secured, in order to make the seeming country landscape readily and comfortably available for all. Each person, young and old, should have the greatest liberty to enjoy himself in a park that is compatible with the enjoyment of neighbors. This somewhat trite aphorism does undoubtedly apply with force to the life and actions of all citizens, but to none does it apply with greater force than to those who occupy our parks.

In building a park, as in building a country place, the truly sympathetic and artistic designer will seek to enter into the spirit of the natural advantages of the region, and to loyally stand by them, never faltering in his determination to resist, as far as lies in his power, all mistaken attempts to make improvements that are really injuries to its characteristic and native charm.

This will mean to retain, in fact, as far as can be made to appear reasonable, all masses of woodlands, valleys and hills, and pools, and streams of water, so that they will be neither cut down or filled up, except for the introduction of such evidently necessary features as roads and buildings that cannot be avoided. Rather is it wise to turn the attention to enriching and perfecting the beauties already evident, by planting, fertilizing, and cultivating on the simple lines indicated by existing conditions, and, in addition, by taking care to always relegate such evidently artificial-looking designs as flower-gardens and colored foliage-beds to entire seclusion behind walls or hedges.

The constant endeavor, in a word, should be to help Nature, and give her free and unrestrained license to develop her native charms as far as the circumstances and comfort of the human element will permit. There should be no attempt to deceive with meretricious rockwork, for example, but everywhere should appear continual suggestions of country.

These remarks should be made to apply to parkways, as well as parks, for there is nothing in a park that is essentially different from a parkway, the latter being simply a high road connecting two parks, and bordered by relatively narrow strips of land on either side, treated as park territory.

CHURCH-YARDS AND CEMETERIES

THE yards of churches, and inclosures known as cemeteries, that have been set aside for the burial of the dead, seem to plead to us for the retention and continuance of all things that will make for quiet dignity and peace and the lingering maintenance of tender memories. They seem to ask us, above all, for absolute simplicity pervading everything, so that no single jar may disturb that quiet and brooding of the soul that should dwell, at least for a time, in the minds of men in the presence of the burial ground. In such places, in church-yards particularly, we seem to desire much smooth, soft, green sward, which rests and unconsciously satisfies the eye, so that only a few flowers, and those of modest and unobtrusive kinds, are needed in addition to it. Many shrubs dotted about will, we feel, jar on that sense of dignity and quiet which we naturally expect in God's Acre.

The trees, for trees there must be for shade, and the feeling of beneficence and protection they suggest, belong on the outskirts of the church-yard, or along the road, and in the park-like reservations of the cemetery. On the boundary line of church-yards and cemeteries may come the fences and walls, or it may well be that the beauty of peace and quietness will be better fostered by the absence of fences, and yet the fence, or wall, seems good to have for the purpose of barring out the desecrating feet of indifferent persons, or the mischief-doing, wandering animals. If fences and walls are used, they may be covered with thick-growing vines, remarkable for their green leaves rather than for flowers, such as the English ivy on the north side of buildings, and the two ampelopsises, namely, the Virginia creeper and the Japanese ivy, also evergreen honeysuckle and the akebia quinata, while the more showy flowering clematises, trumpet-creepers, wistarias, and running roses seem almost too vivid and brilliant in coloring for the even tone and quiet temper that we would naturally consider native to the region.

All trees are not fitted for these retired spots. American and English elms and most kinds of oaks, beeches, ashes, and yews, and dark, Oriental spruces and white pines, but not the weeping willow, or any weeping trees, because they always seem to the author to be making a travesty on melancholy; all the others have a dignity and restfulness of demeanor that comport well with all proper church-yard influences. In the same way shrubs, of which there are kinds suitable for the cemetery, should, in their proper place along the fences, show sober coloring with few conspicuous flowers, and among the shrubs suited for this purpose are the horn-beam, the bushy forms of dogwood, cornus alba, and C. sericea, privets, spicewood, lonicera frangratissima, philadelphus, rhodotypus kerrioides, symphoricarpus vulgaris, viburnum sieboldii, viburnum lantana, and viburnum pyrifolium, and the white-flowering dogwood, cornus florida.

All this may seem to the reader a little fanciful, but

when he learns to approach the decoration of his grounds as he would the canvas of the picture—and are not the church-yard and burial plot most truly his grounds?—he will come to feel very much as the author does in relation to the use of different trees, shrubs, and vines, and not smile indulgently at what he may at first think an over-refinement of sentiment.

Continuing, therefore, the same line of thought, there are, for instance, no kinds of vines that suit the gray stone or red brick walls of a church so well as the sober English ivy, and the broad, smooth, dark-green Japan ivy, yet care should be taken to keep the church from being covered entirely with encroaching tendrils, for half the charm of the climber lies in the contrast it makes with the color of the stone or brick of the building.

Feeling thus profoundly the inharmonious influence in the church-yard of certain kinds of trees and shrubs, and above all of showy flowers borne by herbaceous or climbing plants, we would naturally fail to contemplate with any degree of satisfaction pretentious, cumbersome monuments in church-yards and cemeteries. When hoary with age, half-falling or partially broken, they jar on us less, but no broken or decayed thing can be really restful or peaceful in its influence, any more than the mere expensiveness of the glaringly ugly new monument or tombstone, that speaks of living prosperity in its worst aspects, can give rise to that peaceful melancholy so much to be desired for the associations that should dwell about the burying-ground. It is fortunate, therefore, that fashionable taste for tombstones and monuments of the more vulgar kind has seemed for some time to be waning, and it is a growth of the sentiment that dislikes such things which has induced many persons to advocate, with success, the park cemetery, where every effort is made to do away with huge or ornate tombstones, and particularly the so-called ornamental fences that are always objectionable, and, if possible, to induce lot-owners to build their graves level with the ground, keeping plain tombstones, like those of the colonial period, for record.

There is a gratifying improvement in this respect in most of the cemeteries of the country, and a particularly gratifying development of the park idea, whereby large areas of land are set aside for greensward, trees, and shrubs, which can be chosen with due respect to the sober character that should mark all parts of the cemetery. The joyous element of children romping and playing their games should naturally be banished; but how pleasant and grateful, in a quiet, comforting way, these park-like spaces can be made, many can testify after much walking on sad errands. Indeed, our cemeteries are becoming, in much the same way as our parks and home grounds, places of resort where soul and body will be rested and refreshed.

SEASIDE LAWNS

THE theory of arrangement of seaside places need not differ materially from that of other home grounds, except that, as far as possible, the sea should be made visible from the windows of every living-room in the house; and the roads, if it be practicable, should arrive from the land side. The main thing to be really studied is the preparation of the soil and the selection of plants that should be used, for not all trees and shrubs, by any means, will thrive in even secluded places on the shore.

The chief difficulty one usually encounters at the seashore is the poor, sandy nature of the soil, and consequently it is generally necessary to bring strong loam from a considerable distance; but whatever the distance, we should bring the soil without fail, for on it depends the eventual success of all plantations, and it will be found that a covering of two feet of mold will not be unnecessarily deep to secure satisfactory results.

Blue-grass seed should be sown in liberal quantities on such soil, and plenty of sprinkling applied at once, if the rainfall should not be abundant. With plenty of water and a good top-dressing of mold, excellent lawns can be secured on dry, sandy beaches. The same rule naturally extends to the use of strong loam in the holes where the trees and shrubs are to be planted.

Concerning the choice of trees and shrubs for a seaside lawn, it will prove better to limit one's self to the few kinds that are well known as having the vigor to resist successfully the winds and salt air of the seashore than to fail altogether in attempting to use too many shrubs and most evergreens, that generally behave badly in such regions. A few varieties of pines are exceptions.

Among deciduous trees, there are several that do well on the seashore, and notable among these we find the Oriental plane tree, which is vigorous, well furnished with foliage, and suited to resist the strongest sea-breezes. In the same class will come the honey locust, and the picturesque and always valuable wild cherry, cerasus serotina.

The catalpa has a vigorous habit that suits the sea-shore, and the gingko, salisburia adiantifolia, is also hardy and enduring in similar localities, while the rhus cotinus and R. osbeckii also do well in the salt air, but two of the most valuable trees of this kind will be found to be the golden-barked willow and the laurel-leaved willow. The willows generally are valuable on the seashore, and the same may be said of the hardy and picturesque native American thorns, cratægus crus-galli, etc.

Some of the best shrubs for the seashore are the privets, the bush honeysuckles, lonicera fragrantissima, the different kinds of philadelphus, rhamnus catharticus, the sea buckthorn, myrica cerifera, the tamarisks, the elæagnuses, the althæas, the hydrangeas, and several viburnums, notably V. prunifolium and V. opulus.

There are many herbaceous plants that will seem

almost indispensable when they are seen doing well in the neighborhood of the sea, and the honeysuckles and Virginia creepers we would certainly find that we could not spare. Some of the best herbaceous plants for the seashore are the coreopsis lanceolata, the eulalia japonica, the different kinds of sun-flowers, the irises, particularly iris pseudoacoris, garden phloxes, hollyhocks, hibiscus californicus and H. moscheutus, and the marshmallow, alva alcæa, asclepias or milk-weeds, statice, pyrethrum, or chrysanthemum uliginosum, double and single silphiums, and helenium autumnale. These plants constitute a collection of trees and shrubs and perennials that do well at the seashore, and though there are others, there are but a few others that will do nearly as well.

Before leaving this subject, the author will risk the chance of making himself wearisome by reiterating, in the most emphatic manner, his advice in regard to the importance of using abundant quantities of strong, rich loam on the sandy soil of the seashore, and of applying large amounts of water when the rainfall is insufficient. In that way only can successful lawns and plantations be secured under the stress of sea-breezes and the difficult conditions of beach territory.

On rocky shores the problem remains much the same as on sandy shores, because disintegrated rock is apt to constitute a chief part of the meager soil between the stones, and although such soil, especially if the region be more or less wooded, is not likely to be so poor and unfertile as that of pure sand beaches, yet the bleak winds and sea-air make the growth of trees and shrubs difficult, except by the use of such vigorous species as are mentioned in this chapter.

The value of the abundant use of water and additional rich soil for rocky seashores cannot be overestimated. It is really wonderful to see how much can be accomplished by diligence and skill in lawn-planting on these apparently barren beaches by careful culture, use of water and fresh soil, and by planting skilfully just the right trees, shrubs, and herbaceous plants. Unless the attempt is made too near the sea, or in a specially bleak place, there is more hope of success here than on most pure sand beaches.

CITY AND VILLAGE SQUARES

ALTHOUGH there is a great diversity of size and appearance between the small city square, or triangle, and the great urban park, the fundamental principles governing the designs of each are the same. The larger and smaller spaces alike demand a due consideration of the environment and of the situation. Failure in either case lies in doing too much, or too little, and in doing that which is not appropriate and rational, and thus missing the proper adjustment of the means to the end.

Yet there are some radical differences between city squares and large parks. In cities, where the only places of gathering for the crowds are in open places, and where the play-ground of children is the street, it becomes naturally important to reserve abundant open spaces of gravel or asphalt for seats for grown-up people, and room for the romping of little ones. If there should be more than an acre in a city square, there may be found room for shelter, music and refreshment stands, and these structures should be designed in the simplest, most unostentatious manner possible, so that the square, or small park, may retain its proper and original character of a combination of trees, grass, and flowers, to which all architectural devices should be subordinated and kept entirely tributary.

While convenience and ease should always receive due consideration, it must be remembered that what, after all, constitutes the real square is the grass and shrubs

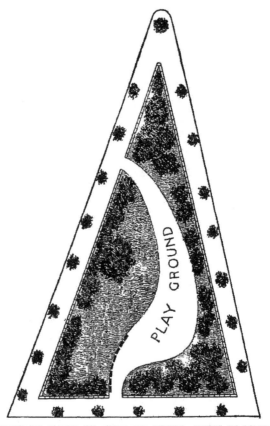

SMALL TRIANGULAR PARK IN CITY OR TOWN, WITH PLAYGROUND AND WALK. (CANAL STREET PARK, N. Y.)

and shade trees, and that without these we cannot imagine reasons for a square being improved as a park. It may be a promenade, a court-yard, or a number of other things that are pure architecture, or anything one pleases, but these are surely not parks. That is why we should always bear in mind that every building constructed in a square destroys the park value of the space it occupies, and detracts in just so much from the true esthetic and essential value of the park. Therefore, as I have said before, let there be but a few and necessary constructions, and those of the simplest and most rustic character. The presence of surrounding houses is simply the environment of the park, and that environment is also, and very properly, a part of the problem to be accepted and accounted for in the design.

It is evident that one would not expect to find a bit of tangled woodland glade amid the architectural environment of a city square. Neither would one expect nothing but stone-paved footways, and series of balustrades, steps, and columns. The way which would commend itself most readily to sensible people would be the arrangement, first of all, of considerable level stretches of smooth, green turf, with groups of shrubs and trees, so arranged as to display their individual charms effectively, and at the same time managing, in an artful combination, to vaguely suggest some such appropriate and beautiful effect as that of a copse in a meadow and a lane in a country home.

The design of a city square should also, invariably, take into consideration the practical features that will be required to protect the greensward and shrubs, and afford convenient accommodations for promenades and playing grounds. Boundary fences, if not fences along the paths, are absolute necessities on squares in crowded districts, and where only their presence will preserve, for any considerable time, the beauty of the park. This is averred with full realization of the efficiency of the police under a good government. When we accept the

236 *HOW TO PLAN THE HOME GROUNDS*

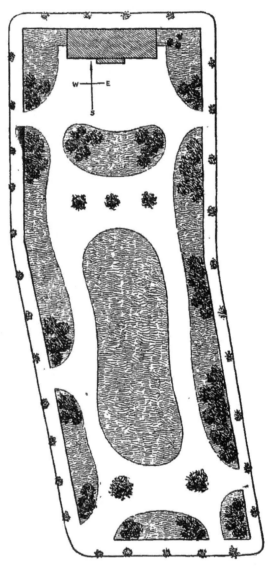

SMALL PARK OF FOUR ACRES FOR CITY OR TOWN. (MULBERRY BEND PARK, FIVE POINTS, N. Y.)

fence, however, we need not necessarily include its hard, disagreeable lines, because they may often be shrouded with vines like honeysuckles or Virginia creepers.

So much space is needed in a small city square for the

convenience and enjoyment of the public, that we will find ourselves limited more than we would like to be, when we are seeking to make the park as beautiful and picturesque as it is in the power of trees, shrubs, and vines to make it. Hence the reason why the allotment for amusement should be kept as much as possible together in one spot with only grass directly around, and the shrubs and flowers fenced in on the boundaries, or on one side, so that they may escape the inevitable destruction that is sure to come when in too close proximity to large crowds of young and old people. A convenient outlet for the high spirits of young folks may be secured in conveniently arranged play-grounds, where there shall be no attempt to create a genuine park, but simply a play-ground with the tree and shrub element almost entirely left out. It is possible, with proper care-takers, to arrange sand-pits for the amusement of little ones in ordinary city squares, but for play-grounds for half-grown boys an isolated area devoted entirely to the one purpose of games is naturally set apart.

To village squares the danger of destruction attendant on the presence of crowds does not apply with the same force, although it may be doubted whether even there trees, shrubs, and grass would not be benefited by the protection of a fence of some kind. However, simplicity and openness are attributes that, in a village square, or green, cannot be dispensed with. A piece of greensward, with a few spreading umbrageous trees, will make an ideal village green, if the trees are set far enough apart, fifty feet or more, to give them plenty of chance to properly develop. When the village green needs protection of some sort, a very low stone wall, covered with vines, will make an attractive boundary, or a wire fence may be used, and covered in the same way.

The ornamentation of village greens with classic fountains, statues, and memorial shafts is, we may be allowed to say, with all due respect to the benevolent and patriotic motives that frequently inspire their erection, to be generally deprecated. While one would not exactly select a pool or stream for the artificial ornamentation of a more or less formal village green, yet it may readily come to exist naturally in a square, in which case a fountain basin would be in order, provided no elaborate-looking designs in marble or bronze are used, but if fountains must be used, much rather should one employ the beautiful single spray of crystalline water, or a cluster of sprays like those to be seen in the contrivance known as the geyser fountain.

With the improvement of village greens should go the proper shade and adornment of the highway bordering it. It is best, as I have stated, that shade trees should be set out at intervals of fifty feet, and the dwelling-houses should be set back as far from the road as circumstances will permit, so as to further extend the openness of the territory.

On a village green, paths should be few in number, open space or spaces should be left for seats and the gathering of people, and, above all things, plenty of shade should be fostered. Straight walks are admissible, if not often advisable, only they should not make acute angles with each other, to the destruction of beauty and vegetation, but where a long curve can be given to a path, without appreciably detracting from its directness, it is better to employ it, and its effect is sure to be most attractive.

On a village green, rock-work seems specially out of place, although if a great natural boulder is found within its confines, it would be well to keep it, but a heap of stones we need view only as rubbish that should be carted away.

RAILROAD STATION GROUNDS

THERE are, we are forced to remember, many meeting places in life where the accidents of travel, or the natural delays of miscalculated time, oblige people to linger in a frame of mind that is almost preternaturally disposed to complain of surroundings that appear doubly uninteresting, for the reason that their contemplation is forced and compulsory; and it is highly probable that the precision, promptitude, and rapidity of action generally associated with a railroad will account somewhat for the special sense of boredom with which a long wait at a railroad station is generally contemplated.

In view of this natural condition of the waiting public at railroad stations, it is no wonder that the minds of railroad managers have been turned for a long time toward the development of the convenience and attractiveness of all railway stopping-places, for the contemplation of a weedy, cinder-strewn yard, and a gullied bank with a freight-car or two standing on the rails, does certainly not conduce to cheerfulness of soul or resignation to enforced delay.

Recognizing that money expended in such improvements will be always profitable, many, if not most railroads, have expended increasingly large sums of money, year after year, in building handsome stations, with convenient roads of approach, and more or less satisfactory plant decoration; but, unfortunately here, as in other cases of park and garden undertakings, the excellence and artistic value of the plant-work seems to lag sadly behind that of the architecture. The best architects in the country have long been accustomed to put forth, under the spur of competition, their best efforts to design the most convenient and beautiful station buildings, while, on the other hand, with very few exceptions, the decoration of the grounds around railroad stations, if attended to at all, has been left to be developed by skill that cannot be said to be either artistic or comprehensive in its scope.

There have been, without doubt, notable improvements in this respect accomplished during the last few years, an excellent illustration of which may be seen on the Boston and Albany road, but it still remains, unfortunately, the practice on most roads to set out a few coleuses, geraniums, and cannas, and there feel that the necessity of the occasion stops; whereas it must become evident to those who will give due consideration to the subject that in the scheme of such improvements the geranium type of plant should usually take a small, and never a dominant, part. Indeed, it would seem natural, when we come to consider it, to arrange station grounds in the same comprehensive way that we would our small parks or private grounds, for all would concede that home comforts and attractions would prove specially agreeable and solacing, both inside and outside the station.

In following out, therefore, this idea, we would have, above all, in such grounds, permanent plantations of trees

16

and shrubs, properly arranged with regard to a park-like effect. There might be bits of color introduced by the use of bedding, but the dominant and permanent idea would be arboreal, and fitted for enjoyment all the year round. It is a little strange that a wider view of the subject has not been more generally entertained, and designs worked out for stations that will include all the possible beauties of the park and lawn, whether trees and shrubs, evergreen and deciduous, or herbaceous, and bedding-plants. Plant for plant, it will be found that the average cost of these different kinds of material does not seriously differ; that is, many of the best shrubs and trees can be bought as cheaply as cannas and geraniums, and the trees and shrubs need no replacing year after year, as the cannas and geraniums do. This would evidently reduce the cost of the maintenance of station grounds greatly, as compared with the expense of an exclusive system, carried out yearly, of decoration with only bedding-plants.

It seems to the author that an important reason why we do not find better systems established everywhere in the improvement of station grounds is because there is generally a lack of method in formulating the designs that are to be used. If a station is to be built, an architect always prepares a plan for it, but in the case of the grounds, this is usually done by any one, and consequently a haphazard and more or less inharmonious result is pretty sure to follow.

It is certain that if a plan of walks, roads, turns for carriages, and open bits of lawn, with plantations of trees, shrubs, and flowers, were always prepared beforehand, greater beauty of park-like effect would result, and during a considerable period of years, on account of the lack of necessity for changes in a well-thought-out design, there would be much less, instead of greater, expense.

Let us turn, in further consideration of the subject, to the accompanying plan of a station on one of the main railroad lines of the country, and see how some of the details of the work should be carried out. It will be readily seen that the stopping-place in question is one of considerable importance, and should, therefore, furnish abundant open graveled space for the rapid gathering

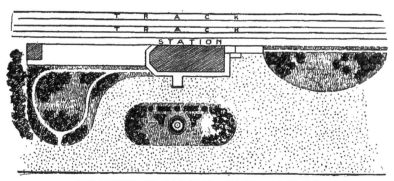

TREATMENT OF RAILROAD STATION GROUNDS

and leaving of carriages and other vehicles. In the center, in front of the station door, however, there has been designed, in order to relieve the general bareness of the open graveled space, an oval or oblong lawn of eighty feet in diameter, in which a fountain basin is located of unassuming character, having no sculpturesque accessories, but simply an abundant spray of water. In this basin a few water-lilies are to be grown, and around its outer edge a small grouping of brilliant-looking bedding-plants, acalyphas, geraniums, and alternantheras, is arranged in varied and well-contrasted masses. This is intended to be the only bit of brilliant leaf-color made by bedding-plants to be found on the place, and as it adjoins the formal lines of the fountain basin, it is in entirely good taste, and, from its limited size, comparatively inexpensive. On the other hand, it is, perhaps, the least important feature of the place, and would be missed less on account of its ephemeral character than either the grass, the trees, and shrubs, or the water of the fountain. On each side of the station, and bordering the railroad, considerable areas of green lawn extend for about one hundred feet either way, giving space for an abundant display of grass, the most inviting object the eye can rest on at a railroad stopping-place.

On the borders of the lawn are disposed, in irregular groups, with a few trees intermingled, various hardy and vigorous deciduous shrubs, that afford attractive leaves and beautiful flowers at various seasons of the year, according to the habits of the different kinds, which include dogwoods, viburnums, spiræa opulifolia, S. thunbergii, philadelphuses, privet ibota, forsythias, berberries, bush honeysuckles, and the Japan rose, rosa rugosa.

The deciduous trees consist of such kinds as elms, lindens, maples, tulips, American ashes, and Oriental plane trees, and are planted fifty feet apart, so as to give them a park-like effect. In this way a varied effect of bark, branches, and foliage is obtained throughout nearly the whole year.

The writer desires to point out that one feature which usually accompanies the railroad station, and that does not happen to exist in the plan we have been considering, is the steep slope or bank that is frequently made by a railroad cut in the front or back of the station. Such a place affords a most excellent opportunity to plant out a woody-looking growth, in every way natural, of run-

RAILROAD STATION GROUNDS 245

ning vines like honeysuckles, roses, etc., and low shrubs with interesting foliage. On some banks larger shrubs like dogwoods, and small trees like birches, may be used with excellent effect, and make a most favorable exhibition of their charms, which could hardly find, in connection with low shrubs and vines, a more fortunate position.

Another happy accessory of such places would be found in the many hardy herbaceous flowering plants: lilies of the valley, violets, phloxes, irises, etc., that could be planted effectively in the immediate neighborhood of the shrubberies, and in this way colonies of hardy, permanent flowers could be established, the bloom of which would afford delight to lingering passengers during nearly all seasons of the year.

There need be scarcely any limit to the variety of chances to use trees properly without injuring the shrubs associated with them, and without producing too much shade to be agreeable for the occupants of the station. For the purpose of improving the station grounds, a few specimens of white pines and other evergreens may be used, but it is not wise to undertake to employ many evergreens, in view of the fact that they are specially liable to injury by storms and insects and variation of temperature, from the effects of which they are slow to recover. Two or three paths are arranged to wind about these park-like areas, and at their junctions, or ends, it is intended that seats shall be placed to permit the passenger to further relieve his hours of enforced leisure. Wherever fences are necessary for safety, they are to be made of solid wire, or iron, in some form, and covered with the most attractive climbers: clematis paniculata; ampelopsis quinquefolia, Virginia creeper, and lonicera halleana, Japan honeysuckle, and the same vines with the Japan ivy, ampelopsis tricuspidata, should be planted so as to cover a large portion of the walls of the station.

The importance of the maintenance of the improvements of the grounds of the railroad station should not be overlooked, although it need not be made more onerous than the painting and cleaning of the interior of the station. If an emulation and love for helping the plants to thrive can be developed among the station agents and other employees that have to do with them, the extent to which the beauty of station grounds may be improved can hardly be estimated or realized. In proof of this statement it is only necessary to take the trouble to look at the little door-yards of the cottages of the workingmen, both in England and America, where a rivalry and enthusiasm have sprung up for horticultural improvements of all kinds.

INDEX

A

American trees and shrubs, 107.
Aquatic plants, best method of growing, 138.
Aquatic plants, best location of, 140.

B

Bedding plants, proper location of beds of, 146.
Bedding plants, skyline of, 149.
Bedding plants, special arrangement of individual, 147, 148.
Bedding plants, value of, 145.
Breezes, 11.
Bridges, location of, 192, 193.

C

Church-yards, inharmonious plants in, 227.
Church-yards, use of fences and walls in, 226.
Church-yards, use of trees in, 225, 226.
Church-yards, use of shrubs in, 226.
Church-yards, use of vines in, 227.
Church-yards, value of greensward for, 225.
Cemeteries, waning taste for showy tombstones and monuments in, 228.
Cemeteries, park-like effects in, 228.
Contour map, 25.

D

Drying-ground, 13.

E

Evergreens, best method of transplanting, 123.
Evergreens, difficulty of transplanting, 123.
Evergreens, value of, 121.

F

Fence, picket, 188.
Fence, value of, 187.
Fence, wire, 181.
Foot-paths, 39.
Front door, 10, 11.

G

Garden beds, 61, 64, 65.
Garden boundaries, 61.
Garden, Colonial, 55, 56.
Garden flowers, 57, 58.
Garden, location of, 57.
Garden propagating houses, 59, 60.
Garden, relative antiquity of, 52, 53.

INDEX

Garden walks, 61.
Grades of territory, 177, 178.
Grading lawn, 48.
Greens, village, fountains for, 238.
Greens, village, statues for, 238.

H

Hedge rows, 188, 189.
Herbaceous plants, definition of, 129.
Herbaceous plants, proper location of, 130.
Herbaceous plants, use of large colonies of, 131, 132.

L

Lawn, fertilizing of, 49.
Lawn mowing, 51.
Lawn, seaside, choice of trees and shrubs for, 230.
Lawn, seaside, rich soil for top dressing, 229.
Lawn, seaside, rocky, 231.
Lawn, seaside, seeding, 229.
Lawn, seaside, use of water on, 229, 231, 232.
Lawn seeding, 50, 51.
Level lots, 5, 176.
Lodge, location of, 177.

P

Parks, improper objects of, 223.
Parks, principles of the arrangement of, 219, 220.
Parks, proper objects of, 221.
Parks, resemblance of parkways to, 224.
Parks, suggestion of country in, 222.
Plantations, skyline of, 82, 83.
Planting, best method of, 87, 88.
Planting, use of mold for, 88.
Plants, best method of using bedding, 90, 91.
Plants, group relations of, 89, 90.
Plants, preparation of beds for, 93.

Plants, proper depth for setting out, 87.
Plants, proper distance apart of, 91, 92, 93.
Plants, proper selection of, 84, 85.
Plants, single specimens of, 80.
Plants, small number of satisfactory, 80.
Plants, use of one kind of, 82.
Pruning, injurious effects of bad, 98, 99.

R

Railroad stations, encouragement of employees to look after grounds of, 246.
Railroad stations, hardy trees, shrubs, and vines for, 241.
Railroad stations, need of planting plans of, 242.
Railroad stations, use of bedding plants for, 241, 242, 243, 244.
Rhododendrons, hardy kinds of, 127.
Rhododendrons, value of rich, yellow soil for, 127.
Road construction, 28, 31.
Road cost, 26.
Road crown, 36.
Road curves, 25.
Road depth, 37.
Road drainage, 32, 33.
Road foundation, 36, 38.
Road grades, 27, 32.
Road gutters, 33, 34, 35.
Road limit, 24.
Road line, 19, 23, 25.
Road maintenance, 40, 41, 42.
Road rolling, 37, 38.
Road specifications, 31
Road turn, 24.
Road waste, disposal of, 37.
Road width, 35.
Roads of gravel, 37.
Roads, shell, 39.
Rock, conservative use of, 173.
Rock copings, 168.
Rock foot and carriage bridges. 170, 171.

INDEX

Rock steps, 168.
Rock, water against, 171.
Rocks, difficulty of setting, 166.
Rocks, principle on which to use more or less, 168.
Rocks, proper grouping of, 165, 166.

S

Shrubs, good all-round, 116.
Squares, city, limit of buildings for, 235.
Squares, city, playgrounds for, 233, 237.
Squares, city, protection of, 235.
Squares, city, sculpture in, 238.
Squares, city, simple treatment of, 234, 235.
Stables, 14, 15.
Summer-houses, 182.
Summer-houses, construction of, 190.
Summer-houses, designing, 191.
Summer-houses, location of, 192.
Summer-houses, use of vines on, 191.

T

Terrace decoration, 74.
Terrace, effect of, 45, 67, 68, 69.
Terrace hedges, 74, 75.
Terrace limits, 73.
Terrace proportion, 77.
Terrace steps, 78.
Terrace turf, 75, 76.
Terrace walks, 69.

Transplanting large trees, 8.
Trees, location of, 47.

V

Vines, distance apart, 180.
Vines, precaution necessary in training, 191, 192.

W

Walls, privacy of, 187.
Walls, stone and brick, value of, 185.
Walls, treatment of, 185, 186.
Water, difficulties of employment of, 151, 152.
Water, good example of use of, 154, 157, 158.
Water, proper use of, 153.
Water, treatment of shores of, 153.
Woodlands, 8.
Woods, cleaning up, 160.
Woods, cultivation of, 161.
Woods, imitation of natural, 159, 162, 163.
Woods, paths in, 161.
Woods, pruning, 159, 160.
Woods, renewing good soil in, 160.
Woods, use of wild flowers in, 160.
Woods, uselessness of planting fresh trees in, 160.
Woods, value of mulching with plenty of leaves in, 161.
Woods, wandering cattle and, 161.

CPSIA information can be obtained
at www.ICGtesting.com
Printed in the USA
LVHW022209170323
741893LV00032B/1318

9 781015 366671